COMMON GROUND

COMMON GROUND

HOW THE CRISIS OF THE EARTH IS SAVING US FROM OUR ILLUSION OF SEPARATION

Lessons in resistance and solidarity from the movement to protect the Earth

EILEEN FLANAGAN

Foreword by
KUMI NAIDOO

SEVEN STORIES PRESS
NEW YORK • OAKLAND • LONDON

Seven Stories Press
140 Watts Street
New York, NY 10013
www.sevenstories.com

Library of Congress Cataloging-in-Publication Data is on file.

ISBN: 978-1-64421-478-7 (paperback)
ISBN: 978-1-64421-479-4 (ebook)

College professors and high school and middle school teachers may order free examination copies of Seven Stories Press titles. Visit https://www.sevenstories.com/pg/resources-academics or email academic@sevenstories.com.

Printed in the United States of America

9 8 7 6 5 4 3 2 1

CONTENTS

Part I: What Is Blocking Change?

Part II: The Cost of Separation

Part III: The Great Turning

Part IV: One World

FOREWORD
BY KUMI NAIDOO

When I'm asked to speak to organizations around the world about the climate emergency, I often reflect on the communication challenge we face. We must be brutally honest about the scale of the crisis and the powerful forces unashamedly resisting the rapid change we need. The importance of communicating this urgency in a way that inspires, energizes, and activates people rather than depressing and disempowering them is of utmost importance. *Common Ground* achieves this balance. By centering the stories of activists on the front lines, Eileen Flanagan conveys the inevitable growing power of people working together along with the grave stakes that motivate them.

Unlike many other books on climate change, *Common Ground* tackles the issue of power, particularly the way power holders maintain control by deliberately and effectively dividing people. I experienced this firsthand as a young activist in apartheid South Africa, a dehumanizing system that pitted Black, white, Indian, and mixed-race people against each other. In my own Indian community, I saw how offering one group a taste of privilege could undermine solidarity and distract people from the broader issues that were hurting nearly everyone. Sometimes our organizations overcame these attempts to divide and rule, but often we missed opportunities to build a broader and stronger movement by staying in our silos, whether they were silos of race, class, or political ideology. In subsequent decades, as executive director of Greenpeace and then of Amnesty International, I have seen many other examples of divide and conquer and am convinced that grassroots movements are stronger when we resist this trap.

The stories featured in *Common Ground* powerfully illustrate how the fossil fuel industry benefits from racial and class segregation, as well as other divisions within our movements. While frankly acknowledging that the impacts of environmental and climate destruction are grossly unequal, the book highlights that we ultimately all have a stake in protecting the Earth, our shared home. Without sugarcoating the unequal risks of activism—within and between countries—Eileen offers different models for people to work together. Grounded in her Quaker spiritual tradition, she frames this germinating solidarity as a potentially positive result of the environmental breakdown we are experiencing.

I first met Eileen in 1988 when I was a young South African in exile. I was surprised by a white person who cared about racial justice as much as she did. She had come back from a trip to Palestine and shared the painful oppression people were experiencing. I also found it laudable that she always used both sides of every piece of paper and avoided using plastic bags, when most Americans and most people globally were more wasteful. Years later, she visited South Africa and Botswana, where she had served in the Peace Corps, and heard firsthand how those countries were disproportionately impacted by the carbon emissions produced in her own country. Eileen recognized that she had a role in challenging powerholders at home, and like many in the climate justice movement, she started engaging in civil disobedience.

When I met Eileen, she had a "Think Globally, Act Locally" bumper sticker on her used Honda Civic. Although I firmly believe that we also need people negotiating and building solidarity on the global level, it's appropriate that most of *Common Ground* focuses on the United States, which has a disproportionate impact on the rest of the planet. It's also fitting that the last chapter takes on the asset manager Vanguard, the world's largest investor in fossil fuels. We do not have time right now to fight every coal, oil, and gas project and every deforestation company. We have to follow the money and shut the flow of capital at its source. Whether you live next to an oil refinery or are inadvertently invested in one, *Common Ground* shows how our fates are intertwined, and how we need love and courage to build a better world for all.

INTRODUCTION

"We are all connected," said Jacqueline Thomas as tens of thousands of people crowded together at a freezing climate rally in Washington, DC. Chief and elder of the Saik'uz people, Thomas was flanked by other Indigenous women (called First Nations in Canada). She spoke about an alliance of many First Nations who were resisting a pipeline designed to pump heavy tar sands oil to the western Canadian coast so it could be sold overseas. Those relatives who lived next to the tar sands were already dying of mysterious brain cancers. Thomas pointed out that ranchers and Indigenous people rely on the same water, which could be contaminated by carcinogens if the pipeline spilled. "Enbridge has really brought our communities together in Canada," she said, thanking the pipeline company, which brought laughter from the shivering crowd. "Never in my life have I seen white and Native work together before now."[1]

The 2013 Forward on Climate rally included many famous environmentalists, but Chief Thomas's speech was the one that stayed with me. I was inspired by the alliance she described, which ultimately succeeded in stopping the multibillion-dollar pipeline project. I had been hearing other examples of people joining forces across historic differences to protect the water, air, and climate. As an activist and writer, I set out to learn how we could make such collaborations stronger and more common. Over several years, many courageous people from diverse backgrounds shared their wisdom about what works. Through my travels to frontline communities, I also got a clearer view of why such efforts have often failed to prevent new fossil fuel projects. They turned out to be some of the same forces that facilitated Donald Trump's reelection in 2024.

Having listened to oil and gas leaders complain to each other about taxes and regulations, I wasn't surprised to learn that they spent at least $75 million to support Trump's reelection. The amount is many times greater if you count hidden money and contributions from industries deeply entwined with fossil fuels.[2] Special interests have also played a covert role in fostering the mistrust of government that Trump exploited in his campaign. Dating back decades, conservative think tanks—some funded by fossil fuel barons—have spread the falsehood that government unfairly helps people of color. More recently, the attack has been against long-standing civil servants with divisive messages about race used as justification. These claims are fed by the very same racial stereotypes I found were used to dismiss communities of color that fight for clean air and water in their neighborhoods, which even the first Trump administration acknowledged are disproportionately burdened with pollution.[3]

Immediately after his second inauguration, Trump implemented the fossil fuel industry's agenda even more flagrantly than in his first term. As wildfires raged around Los Angeles in early 2025, Trump announced plans to withdraw the United States from the Paris Agreement on climate change, roll back environmental protections, and fire environmental watchdogs. The administration's denial of climate change was so deep, it planned to cut 84 percent of the staff who help people rebuild after natural disasters, like Hurricane Helene.[4] Meanwhile, government agencies were told to remove climate and weather data that farming and fishing communities rely on.[5] The administration announced funding freezes for wind projects, including in Republican districts. Even the military, which recognizes that rising heat and sea levels will impact their operations, had climate initiatives cut.[6]

If the courts, Congress, or the public don't stop him, Trump's policies will accelerate the chaos caused by climate change while weakening our resilience to it. Those hit first will be those who live closest to polluting facilities, as well as those who live in areas prone to flooding, wildfires, and scorching heat. But ultimately, these policies will hurt everyone. Most Democrats can't figure out how to make this point without saying "I told you so" to Trump voters, at least some of whom they will need to win over if they ever hope to regain a majority.

Trump came to power and plans to govern through a strategy of divide and conquer. Under the cover of undoing anything related to racial diversity, his administration slashed the jobs of those tasked with protecting America's most polluted communities, as if the air over those neighborhoods doesn't spread. He cut a collaboration between Indigenous communities and Earth scientists, as if understanding how to live in balance with the Earth is a special interest.[7] Such divisive tactics are not new, but they are now on full display. Today, we all need the wisdom of those who have long navigated the divide and conquer game even more than when I began research for this book.

IF THERE is a silver lining to this colossal storm, it is that Trump is proving—more clearly than grassroots activists have yet managed—the corrupting role of greed in our political system. Like the Canadian pipeline company that Chief Thomas resisted and thanked, Trump could ultimately bring our communities together—if we seize the opportunity. Those who have lived through natural disasters know that storms can have this effect.

BOUNCING FORWARD

One night, in a packed Philadelphia union hall, I heard several panelists describe how New Yorkers united after Superstorm Sandy ripped apart their city in 2012. From all five boroughs and a wide range of grassroots organizations, ordinary people decided they didn't want to just bounce back after the deadly disaster ripped down power lines and flooded subways. They wanted to "bounce forward"—to use the extensive rebuilding process to make their city better for everyone.

For some, this meant reducing greenhouse gas emissions, which trap the sun's heat and warm the Earth, making disasters like Sandy more frequent, severe, and costly. For others, bouncing forward meant preparing the city for future storms, while addressing long-festering problems that Sandy revealed and exacerbated. Eddie Bautista, the executive director of the New York City Environmental Justice Alliance, said that low-income communities and communities of color had warned about the danger of industrial waste long before Sandy's storm surge unleashed carcinogens in many neighborhoods. Eddie said that future storms could be even worse than Sandy in terms of spreading dangerous chemicals. After all, much of New York's toxic waste was in flood-prone neighborhoods, and climate change was churning it up.

Afro-Latino with a gray beard, Eddie grew up in the 1970s in the historically working-class Red Hook neighborhood, a part of Brooklyn that juts into New York Bay. While Wall Street also flooded after Sandy, Red Hook was hit much harder and with much more chemical exposure. So was Gowanus Canal, an old industrial waterway thick with sewage and the dregs of the early oil, coal, and gas industries.[8] One artist found oily gunk on his skin after trying to save his computer and some of his art from his flooded basement in a gentrifying neighborhood a few blocks from the canal.[9] As Eddie pointed out, inadequate environmental testing prior to the storm meant that people didn't know exactly what health hazards had contaminated their homes and businesses.

"There's a saying that our movement has: 'If you're not at the table, you're probably on the menu,'" Eddie explained when I visited his first-floor Brooklyn office to hear more of this story. "We realized that rather than wait for the government to come up with a rebuilding plan after Sandy, it was incumbent on us to develop our own recovery resiliency framework, present it to government, and then see how well government responded." At their first daylong meeting, two hundred people showed up from all five boroughs of New York as well as Long Island and New Jersey. At round tables, the diverse group generated ideas about how to use the crisis to improve their neighborhoods. One priority was to prevent environmental hazards after future storms. Another was making sure that local people benefited from rebuilding jobs. This was one of many ways that New Yorkers had learned from New Orleans, where many construction jobs went to outsiders after Hurricane Katrina devastated the Gulf Coast seven years before Sandy.

New Yorkers synthesized their recommendations into a proposal, which was endorsed by groups representing youth, labor, lawyers, environmentalists, and neighborhoods.[10] The coalition presented its solutions to city, state, and federal officials, who were developing their own recovery plans. Eddie told me that some community ideas were adopted, such as training and hiring public housing residents for jobs repairing the thirty-five thousand public housing apartments that Sandy damaged. Realizing that climate disasters would exacerbate poverty and other inequalities, New York advocates who dealt with labor, housing, immigration, and the environment started collaborating more than they ever had before the storm.

WHEN GLOBAL climate organizations decided to mobilize a giant march in New York City around the September 2014 United Nations Climate Summit, many local groups joined the effort, recognizing it as a special opportunity to build momentum for change. The march highlighted survivors of Hurricanes Sandy and Katrina, as well as youth, who will bear the brunt of escalating climate chaos. There were New Yorkers trying to prevent fracking, a destructive way of extracting deep gas, which New York State banned a few months later. Other frontline struggles included Ojibwe women resisting a proposed oil pipeline through northern Minnesota and Appalachians working to stop mountaintop removal coal mining. By bringing together those marching to protect the land where they lived and those marching for the global climate, the event mobilized a record-breaking four hundred thousand people for these issues.

Chief Dwaine Perry represented the Lenape, who stewarded the island they called *Mannahatta* for thousands of years before the Dutch landed there, looking for beaver skins to sell in Europe, where the animal had been driven to near extinction. Nearly four centuries after the Dutch renamed the island Manhattan, Perry said that the sunrise ceremony before the People's Climate March was a powerful, eclectic convergence of spiritual people from around the hemisphere. "It was all focused to lift the spirits of human beings and to begin the healing of the earth. I would like to think that my people have dreamed this, and now here it is."[11]

Media was clustered near the march's start at Columbus Circle, where people held giant sunflower signs. People streamed by in traditional dress from countries around the world. I walked nearly four miles past union members, students, scientists, traditional environmental groups, families, and people of faith wearing a collage of T-shirts. There were colorful signs, banners, and floats, including one of Noah's ark. There was singing, chanting, drumming, and a minute of silence for the victims of climate change, followed by a raucous "sounding the alarm." The day was electrifying.

ALTHOUGH I'D recruited people to attend the People's Climate March, I didn't realize its full impact until Eddie Bautista laid out the story for me in his Brooklyn office a few years later. Amid the public pressure that the march generated, Mayor Bill de Blasio announced that New York City would reduce its greenhouse gas emissions 80 percent by 2050. It was the largest city to set such an ambitious goal.

Building on relationships forged by the march, coalition organizing expanded, building political influence. During the first Trump administration, New York State passed a groundbreaking law limiting greenhouse gas emissions and creating tens of thousands of jobs in energy efficiency and solar development. Although the state is unlikely to meet all its ambitious benchmarks, the law has thwarted new gas plants while many other states continued building fossil fuel projects with little to no restraint.[12]

More than a decade after Sandy, New York State continues to lead the nation with innovative climate initiatives. For instance, a 2024 law requires the biggest oil and gas companies to help pay the cost of protecting communities from climate chaos rather than having that cost fall only on taxpayers.[13] Industry and their political allies are fighting back, but the fact that such a bill even got signed into law is a testament to the power of a coalition that now includes hundreds of organizations representing frontline communities, people of faith, labor, elders, and youth.

The natural world continues to provide the motivation for this work. In the years since Sandy, both of my adult children moved to Brooklyn, where people continue to experience weather made more extreme by climate change. While those who live in basement apartments or near toxic waste are among the most vulnerable, even the expensive private school where my daughter teaches had to shelter in place after five days of rain flooded Park Slope, sweeping cars down the street. It was a reminder that we all have a stake in finding ways to work together.

THE CROSSROADS

It's not just about the climate. A million of Earth's species face extinction from a combination of pollution, urban sprawl, industrial agriculture, logging, and other human disruptions. Because species rely on each other for survival, scientists warn of a potential domino effect, threatening whole, interconnected ecosystems.[14] Even if humans were not altering the Earth's atmosphere, we'd be heading toward dangerous ecological limits. The fact that industrial growth has been powered by the burning of fossil fuels only makes the other dangers we face more urgent and potentially catastrophic.

Fossil fuels—coal, oil, and gas—are not the only causes of climate chaos, but they are the biggest source of the world's emissions.[15] There's no way of addressing the crisis without moving away from these energy sources. The carbon dioxide and methane released through their extraction and burning trap the sun's heat, in effect melting glaciers and Arctic ice while warming and expanding oceans. This is what makes hurricanes and cyclones more frequent and severe. It's also why making change quickly is so urgent. Ice reflects sunlight away from the Earth, so its disappearance exacerbates warming; same with the melting of frozen Arctic soil, which releases more greenhouse gases.[16] As with local ecosystems, global systems are interconnected. As a planet, we are speeding toward a tipping point, after which climate chaos will take on a life of its own, reinforcing itself in feedback loops humans won't be able to stop.

If we don't turn away from this path very soon, the result of these feedback loops will play out over the lives of our children and far beyond. Hundreds of low-lying cities could be submerged, from New Orleans and Miami to Jakarta and Mumbai. As warm air holds moisture for longer, areas from Central America to Syria could experience even worse drought. Wildfires could make Australia and parts of the American West uninhabitable. Other places could lose their iconic characteristics. Corn could become hard to grow in Iowa as well as coffee beans in Ethiopia. Ireland could stop being green. Moreover, the famine that drove my Irish ancestors to migrate across the Atlantic could pale in comparison to the scale of famine caused by a changing climate. It's not guaranteed that all these things will happen, but the longer we delay cutting greenhouse gas emissions, the more likely catastrophic predictions become.

ONE PIECE of good news is that many more people are aware of these dangers and are convinced of the need for change, both

within the United States and around the world.[17] Five years after
the unprecedented march that mobilized four hundred thousand people in the streets of New York, a youth-led climate strike
mobilized several million people across 156 countries from Brazil
to Borneo in September 2019.[18] In New York City, young people
chanted, "Sea levels are rising, and so are we." In my hometown,
Philadelphia, I supported the youth march as a marshal and felt
a surge of hope as students of all backgrounds filled the streets,
walking so quickly the elders struggled to keep up. In part because
of the political pressure built by bold youth organizing, the
United States passed legislation that spurred renewable energy,
which was already increasing nationwide and globally much faster
than expected.

In the past thirteen years, I've served as a leader in three climate campaigns, helping to build connections between different
groups. I've committed civil disobedience and been arrested with
Black preachers, Indigenous water protectors, rabbis, and Quakers
like myself. These experiences have made me more convinced than
ever that coming together across our silos and divisions is key to
making progress on climate protection, as well as every other issue
that ails us.

In the face of the current backlash, it's easy to believe that
recent progress doesn't matter. I believe the opposite. New York's
climate law and the rapid growth of renewables still have a positive
impact. Maybe even more significant, over the past fifteen years
people have been learning what helps to build momentum for
change, and what doesn't. It's a crucial time to harvest and apply
those lessons. It's also a good time to learn from earlier movements that navigated the changing tides of progress and backlash.
One lesson from many successful movements is that people need
a vision to guide their way.

A VISION OF POSSIBILITY

My own vision was kindled by Chief Thomas and other Indigenous people. When I heard her speak at the 2013 Forward on Climate rally, I had just written an article about a comparable alliance in South Africa. Black members of the Landless People's Movement and white farmers, whose ancestors had taken African land, had long been at odds. Now they were working together to protect the fragile Karoo desert from fracking. I was moved to hear that they were partly inspired by *Gasland*, a film about fracking in the United States, which showed the flammable water and mysterious health concerns that appeared after undisclosed chemicals were used to extract methane gas. When a fracking spill caused families to be evacuated from a town in my home state of Pennsylvania, South Africans publicized the incident, winning a temporary moratorium on Shell's plans to frack in their country.[19] Meanwhile, learning about the water and food scarcity that climate change was causing in southern Africa, where I had served in the Peace Corps in the 1980s, was part of what motivated me to commit civil disobedience for the first time.

My first arrest was part of the campaign against the Keystone XL Pipeline. It started with the First Nations people who lived near the Alberta tar sands, who reported rare brain cancers after the toxic process of extracting heavy tar sands oil began. Their organizing inspired people in Canada and the United States to oppose the pipeline, intended to pump tar sands to the Gulf of Mexico for refining. Linking Indigenous sovereignty and climate, the campaign mobilized new people and forged groundbreaking coalitions, like the Cowboy-Indian Alliance in Nebraska. It pressured three US presidents, both Democrat and Republican, before Joe Biden canceled the Keystone XL Pipeline's border-crossing permit, essentially killing the project. Although the second Trump administration has threatened to revive it, the main company involved has already sold the pipes and seems to have moved on.[20]

I thought of these coalitions during a 2014 gathering at Haverford College to explore the topic of "Living in Right Relationship," a concept that implies harmony with ourselves, other people, creation, and the Creator. During the opening program, the audience was asked to reflect on our visions of living in right relationship. The speaker before me shared her anxiety about several environmental problems, but no vision. I raised my hand and was handed the mic. As I stood in front of the crowd, I felt I was channeling the words rather than composing them. "It's not about us saving the Earth," I said. "The Earth will survive whether we do or not. It's that the crisis of the Earth is saving us from our illusion of separation."

THE CRISIS *of the Earth is saving us from our illusion of separation.* These words described what I found encouraging about the stories of Indigenous and non-Indigenous people working together. In so many places, colonizers perpetrated tremendous harm in part by convincing themselves that they were separate from and superior to the people who lived on that land before their arrival. They also brought a worldview that said they were separate from the land itself. As a result, they built an economic system that treated people and the planet as disposable. Now, the world is littered with places like Gowanus Canal, where toxic waste was dumped into the water without anyone imagining that the Earth might someday cough it up. The Environmental Protection Agency (EPA) has identified 945 sites where toxic waste is vulnerable to being spread through increasing wildfires, flooding, or rising sea levels. There are likely many more not on the agency's list.[21]

This is just one example of how climate change is pushing us to reexamine modern assumptions about our relationship with the Earth. Nineteenth-century engineers turned a creek surrounded by wetlands into the smelly Gowanus Canal without appreciating how wetlands help to reduce the impact of hurricanes. Now, helping wetlands rebound is part of climate resiliency in New York, Louisiana, and many other places. Similarly, the engineers who built New York's subway system disregarded the old underground water routes, which now reassert their claim after major hurricanes.[22] Such events are drawing more people to recognize the wisdom of Indigenous teachings about living in balance with the Earth. After all, the Lenape and other original stewards of Mannahatta left only mounds of oyster shells as waste.

In my Quaker faith, we often speak about insights coming from Spirit or God. That was how I experienced the message about "the illusion of separation," that the words came from a source of wisdom beyond my own mind. For me and some of the people I interviewed for this book, that spiritual source is the root of our interconnection. I later discovered that diverse spiritual teachers also use the phrase, though without a shared or precise definition. I also came to understand that expressions of unity can backfire when not grounded in a recognition of the great harm perpetrated by the systems that divide us.

THE PAIN OF SEPARATION

I experienced the pain of feeling separate from someone I care about after a 2014 rally to prevent Philadelphia from becoming a hub for fracked gas. I attended with Matthew Armstead, a facilitator and trainer whom I worked with on another local climate campaign. While the crowd was predominantly white, the racially diverse speakers stressed that polluting fossil fuel facilities were likely to be built in low-income communities of color, where asthma rates were already tragically high. I'd heard the message about our illusion of separation only a few months earlier and mentioned to Matthew on the drive home that my favorite speech was the one where a rabbi said, "We are all one."

"I hated that speech," said Matthew, who is Black, queer, and uses the pronoun *they*. Matthew had recently spent three weeks supporting young activists in Ferguson, Missouri after Darren Wilson, a white cop, shot and killed Michael Brown, an unarmed Black teenager. Brown's body was left in the street for hours, igniting national outrage at racist police violence. The cop had just been exonerated when Matthew explained, "We may all be one, but we're not exactly experiencing the world that way at the moment." I felt the truth and importance of their insight. I also felt how my identity and experience as a white person could make my professions of unity fall flat, or worse, add salt to my friend's wound in such a polarized moment.

IN HINDSIGHT, I should have known better. Prior to becoming a climate organizer, I taught university courses on racism and the struggle against colonialism in southern Africa. Having facilitated many conversations on these topics, I knew that the lines used to divide us can be painful, both because of the oppression they enable and because they can make connection more difficult. Old patterns of dominance and exclusion die hard. Building trust is slow and painful, especially when the original injustice has not been rectified. The collaborations I found inspiring were also likely to be challenging.

In my work as an activist, writer, and speaker, I have intentionally started trying to ground myself in the spiritual and ecological truth that we are all one, alongside the painful reality that some people are treated as expendable, whether by politicians, polluting industries, or the police. Climate change is a perfect issue for teaching us not to lose sight of this complexity since it affects everyone in the world, though not equally. A few months after my conversation with Matthew, I was asked to speak at a Quaker climate event, which was advertised with the line, "We're all in this together." I suggested to the audience that we're all in this together like we're all on the *Titanic* together. Some of us have access to lifeboats, but we're all on a ship headed toward catastrophe. We can fight over the lifeboats, or we can work together to steer the ship in a safer direction.

In the decade since these events, it has become painfully clear that some powerful people want us to fight over the lifeboats. Donald Trump is a master of demonizing groups of people for political gain, but he's far from alone in using this strategy. I've seen corporations pit neighborhoods against each other when they want to build a new polluting project. Often, they set those who want good paying jobs against those who want clean air and a stable climate, as if people are for one or the other. The captains of division often succeed by stoking existing fears or prejudices, which is one reason we need to find common ground without denying the different ways we experience the world. These lessons from the climate movement are especially timely, as many people with other urgent concerns look to enlist potential allies and build coalitions that can resist the path of division and destruction.

At a time when religion is being deployed to foster division, it is important to remember that wisdom traditions from around the world have taught that we are part of a greater divine whole and that our happiness is bound up with the well-being of others.

Today, it's never been clearer that these wisdom traditions are right. The old strategy of circling the wagons and caring only about our own is not just spiritually shallow but, as a species, suicidal. Science is reinforcing this understanding, showing us that what happens in New York affects Mumbai, and vice versa. For this reason, I continue to hope that the climate crisis might finally make humans realize how much we depend on each other and the Earth that sustains us. These issues are global, and an attitude of "America first" will never solve them.

A TIME OF PROPHECY

Part of what I find inspiring about Indigenous-led campaigns, like the one against the Enbridge pipeline in Canada, is that they are grounded in a worldview of connection without sugarcoating the long history of harm their communities have experienced. Sherri Mitchell expanded my understanding of these struggles through her profound book *Sacred Instructions: Indigenous Wisdom for Living Spirit-Based Change*. I felt my own spiritual message affirmed when I found that she used the exact same phrase.

"The illusion of separation has prevented us from unifying for millennia, using fear and superstition to create shadows and divisions," writes Mitchell, whose Penobscot name is *Weh'na Ha'mu' Kwasset* (She Who Brings the Light). "We are on the precipice of an evolutionary leap, one that requires us to transcend our differences and integrate into a more harmonized way of being. All of our prophecies speak of this time: the time when the people of the world would begin waking up and unifying for the protection of life." Although she acknowledges that we could miss the opportunity and remain stuck in our illusions, she asserts, "Every person alive today is part of the dream of the ancestors. We are the fulfillment of prophecy."[23]

Mitchell cites several Indigenous prophecies, including one from the nineteenth-century Lakota leader Crazy Horse, who spoke of a time seven generations hence, a time of "broken promises, selfishness, and separations . . . when all the colors of mankind will gather under the sacred Tree of Life and the whole Earth will become one circle again. In that day, there will be those among the Lakota who will carry knowledge and understanding of unity among all living things, and the young white ones will come to those of my people and ask for this wisdom." Many believe this prophecy was fulfilled at the Standing Rock Sioux Reservation during the 2016–2017 resistance to the Dakota Access Pipeline, in the same region where Crazy Horse delivered the prophecy seven generations earlier.[24]

After people in Bismarck objected to it, the pipeline was rerouted to pass through Standing Rock, including through sacred sites and burial grounds. When the tribe's request for a court injunction was rejected, they invited people to come support them in civil resistance. Over a matter of months, the campaign drew Indigenous people from 280 tribes, as well as several thousand non-Indigenous people of all races. They bravely stood their ground against rubber bullets, tear gas, and attack dogs, many crediting prayer for the fact that no one was killed.

"There were so many good people," one Indigenous elder told me of his months at Standing Rock, echoing a sentiment I heard from other participants. Many people, Indigenous and non-Indigenous alike, went home more committed to protecting the water and land around them. Others who heard about the campaign were also inspired, even though the resistance didn't stop the pipeline's completion.

I'm heartened by the example of people who are trying to find common ground. They bolster my hope that being forced to work together for our common future will actually spur us to heal our divisions, repair harms, and find a more just way forward. This vision feels like both a political strategy and a spiritual path. Given how politically polarized the issue of climate change has become, we need to build a much broader movement to have any hope of thwarting the corporations that continue to put their profits over the health of people and the planet. I've become convinced that we won't be able to do that if we're primarily driven by fear.

In many ways, this book is my attempt to answer a query posed by nonviolence trainer Kazu Haga before the 2024 election: "How do we mobilize the power needed to stop injustice while cultivating the love that's necessary to heal it?"[25]

PART I

What Is Blocking Change?

CHAPTER ONE

A CLASH OF WORLDVIEWS

As I took a seat at the 2019 Louisiana Mid-Continent Oil and Gas Association (LMOGA) annual meeting, I glanced nervously at my neighbors and wondered if they could tell I did not belong. I'd bought a new black dress and put on my only pair of high heels. Even though I was donning lipstick and mascara, I felt conspicuously unbusinesslike with a backpack instead of a briefcase and short, graying hair instead of the more time-consuming hairstyles of the industry women. Most of the people gathered at this glitzy New Orleans hotel wore name tags that identified them with companies like Exxon, Shell, and Chevron. Mine simply read "freelance." As a Quaker writer, I had never done research quite like this, embedding myself in hostile territory.

Five months earlier, the world's top scientists issued an urgent report, warning that to avoid all-out climate catastrophe, humanity had to cut in half our use of oil, gas, and coal by the year 2030.[26] I'd come to the conference to listen to the oil and gas industry's side of the story, to better understand why they continued expanding fossil fuels in the face of such dire scientific predictions. What I knew, and other attendees didn't, was that some of my activist friends in New Orleans planned to disrupt the conference on its second day to assert a worldview that prioritizes the care of people and the planet over profits.

FROM THE first panel on, it was clear that profit was the primary concern at the two-day conference. Anne Idsal, regional administrator of the Environmental Protection Agency (EPA), flattered her audience on their innovation and acknowledged their desire for predictable regulations and short permitting processes, so they could "make investments that would make sense for the next twenty years." Chuck Carr Brown, who worked for Exxon for twenty years before becoming Louisiana's top environmental regulator, proclaimed, "You won't have any problem sustaining your profitability. We are going to be supplying natural gas to the rest of the world for the next forty years." Brown went on to jovially warn his former industry colleagues that he might increase their fines, but not by much.

I tried not to show my dismay. While most people I knew were alarmed by the recent climate report, here even the environmental regulators were predicting decades of fossil fuel production. Equally surreal was the way conference speakers avoided any acknowledgment of how climate change was already affecting their industry, let alone the rest of the world. One panelist complained that rapid coastal erosion, coastal highway flooding, and hurricanes were serious challenges to offshore oil drilling. He omitted the fact that his industry was a major contributor to climate change, which exacerbates all three.

The illusion of separation was evident throughout the conference. Oil and gas executives dismissed the health concerns of their fellow Louisianans, whom they only mentioned as potential plaintiffs, when they mentioned them at all. In New Orleans, I had been warned not to drink the tap water because it contains so many pollutants. In Cancer Alley, the infamous petrochemical corridor that stretches from New Orleans to Baton Rouge, I'd smelled the pungent air near oil refineries and heard about the blaring alarms that wake people from their beds when there is an accident. I had seen how close some homes are to chemical plants, which are built near oil and gas facilities to conveniently use their by-products. I had heard people describe the rawness of their skin or throats after chemical exposure. I'd also heard petrochemical industry workers describe their own high cancer rates and the chemical accidents that had killed colleagues.

GOVERNMENT LAXITY has consequences. In 2010, BP's Deepwater Horizon oil rig exploded off the coast of Louisiana, killing eleven workers. The disaster released an estimated 210 million gallons of crude oil into the Gulf of Mexico, wiping out millions of animals, from tuna and oysters to turtles and dolphins.[27] More active government oversight of deep-sea drilling could have prevented the cascade of failures that led to the disaster. Instead, the offshore oil industry received billions of dollars in government subsidies.[28]

Scott Angelle was the lieutenant governor of Louisiana at the time. Downplaying the risks, Angelle led the push to resume deep-sea drilling as quickly as possible. Later, while leading a federal bureau charged with drilling safety, he saved the oil and gas industry at least a billion dollars by rolling back federal safety regulations implemented after the BP disaster.[29] Yet, Angelle was introduced at the LMOGA conference as a great lover of the environment.

"We can enjoy some of the best hunting, camping, fishing on the globe, *and* produce energy," proclaimed Angelle, a proud Cajun, descended from French Canadians who migrated to Louisiana in the eighteenth century. Listening to him in the glitzy Ritz-Carlton ballroom, I remembered the Cajun crawfisherman I'd met, who said that oil and gas pipeline construction was destroying the unique ecosystems they relied on to make a living. In contrast, Angelle and other speakers implied that environmental protection was only about recreation—not the food we eat, the air we breathe, the water we drink, and the climate that makes Earth habitable.

The BP accident could have taught them otherwise if they had paid attention to the connections the disaster revealed. A year after the spill, 80 percent of pelican eggs in Minnesota were found to contain toxins from the oil and its cleanup, which didn't remove the oil so much as sink it by spraying it with other chemicals. This statistic is less surprising when you realize that the Mississippi River, which empties into the Gulf, forms the migratory path for about half of North American birds. Sea turtles were projected to have carried toxins from the Gulf all the way to West Africa. Meanwhile, an estimated one hundred thousand humans along the coast became severely ill from chemical exposure, as well as many of the workers involved in the cleanup. There is no telling how many seafood eaters consumed smaller concentrations of chemicals once Gulf Coast fishing resumed.

Despite my lingering questions about seafood safety nine years after the BP spill, I had enjoyed Louisiana's iconic seafood dishes, from gumbo and jambalaya to charbroiled oysters and étouffée. At the LMOGA happy hour, I couldn't resist the shrimp and raw oysters, which reminded me how easy it is to ignore what we don't want to think about. But listening to oil and gas leaders celebrate their industry in the French Quarter Ritz-Carlton, which flooded during Hurricane Katrina, brought home to me how dangerous denial can be.

DEADLY DISCONNECT

When Katrina hit the Gulf Coast in August 2005 with 175 mph winds, the storm wreaked havoc in several states, especially Louisiana, Mississippi, and Alabama. At the time, I was caring for my dying mother and two elementary school children. My eyes filled with tears when I heard of families who could have gotten out, but who stayed to care for their elders. Others couldn't afford to evacuate because payday hadn't come yet. Some just believed they could ride it out, as they had previous storms. For me and many others, watching the heartbreaking television footage of floating corpses and dramatic rescues, we did not feel separate from the suffering. As I held my young children, it started to sink in: This is what climate catastrophe looks like.

The impact of Katrina was both broadly felt and unequal. Across the region, over 1.5 million people evacuated, of every race, class, and generation. Many lost their employment as well as their homes from the disaster.[30] In bowl-shaped New Orleans, 80 percent of the city flooded, even the Ritz-Carlton. The largest concentration of deaths was in the Lower Ninth Ward, the predominantly Black neighborhood that ninety years earlier was chosen as the site of the Industrial Canal, which provides industry ships a shortcut from Lake Pontchartrain to the Mississippi River. Experts long predicted catastrophe if storm surge breached the canal; still it was not properly reinforced. While many neighbors slept, believing the worst of the storm was over, the canal wall exploded, releasing a torrent so strong it swept homes off their foundations.

It wasn't just the storm or neglected infrastructure that killed people—for days, the city, state, and federal government failed to coordinate help for the fifty thousand people who were stranded in New Orleans's city center. Michael Brown, a wealthy political appointee with no emergency experience, headed the federal relief effort but seemed detached as people died on the streets from neglect. President George W. Bush showed his own disconnect by declaring that "Brownie" was doing a "heck of a job." Ten years later, in an attempt to rehabilitate his reputation, Brown wrote an article for *Politico* saying that "the worst part" of his experience of Katrina was not the dead bodies floating in the streets but being blamed for it all. His biggest regret was not resigning before the storm.[31] Ironically, Brown titled his book on Katrina *Deadly Indifference*, in which he blames everyone else and dismisses allegations of racism.

KATRINA EXPOSED a worldview chasm that would widen over the following two decades. Americans had starkly different beliefs about the role of climate change in causing the storm, the role of racism in causing unequal impacts, and the role of government in addressing such problems. The disaster helped galvanize scientific research on hurricanes, proving that rising sea levels and warming water are making hurricanes more frequent and severe. Yet studies of those who survived Katrina found that among Katrina survivors, "white individuals were substantially less likely than Black individuals to believe in climate change."[32] Another study suggested that the aftermath of Katrina helped some white Americans better understand racial inequality, though others discounted allegations of racism, perhaps because they posed "a threat to the perceived fairness of our system."[33]

I believed in both climate change and racism before the storm, but my understanding of each deepened as I followed the recovery efforts. I was struck by how many decision-makers seemed to fear those in need and how easily people believed rumors of widespread looting and rape, which later investigations found to be greatly exaggerated. National Guard troops aimed their guns at the people they were supposed to be rescuing, which was a clear violation of military code. One famous video shows the coordinator of the military relief operation confronting white soldiers in tanks. "Weapons down, dammit! Put those damn weapons down!" yelled Lieutenant General Russel L. Honoré. They were on a rescue mission, and US citizens were not the enemy.[34]

Lt. Gen. Honoré was not alone in recognizing the humanity of the survivors. Ordinary Louisianans of all races rescued each other from flooding homes. They fed and housed each other. Outsiders also helped. My husband and I donated two plane tickets that we'd been awarded when bumped off a flight. We later met the couple who used the tickets to get home to New Orleans after they were evacuated to Philadelphia. Ours was a small gesture, but one motivated by a belief that we should love our neighbors, wherever they live, and in whatever ways we can.

Malik Rahim, a former Black Panther, cofounded a hugely effective mutual aid group, now called Common Ground Relief. He told a reporter that the tens of thousands of volunteers who arrived to help after Katrina "did more to unite this community than anything. It showed the African Americans here that not all whites were exploiters or racists. And it showed the white folks coming in that all the African Americans here weren't like how the government said. We weren't all criminals."[35] Reading Rahim's account years later reminded me of the spiritual message I'd heard—that the crisis of the Earth is saving us from our illusion of separation. It was encouraging, even if it wasn't true for everyone yet.

BELIEFS MATTER

After Katrina, the multinational corporation Halliburton grabbed up lucrative construction contracts, denying jobs to local contractors who desperately needed the income.[36] Dumping companies charged extortionist fees for removing hurricane debris. Developers scooped up land, driving up New Orleans real estate prices.[37] Simultaneously, groups like Common Ground Relief helped however they could, from tarping roofs and clearing out houses to providing medical care and legal aid.

"Katrina was an extreme version of what goes on in many disasters," Rebecca Solnit observes in *A Paradise Built in Hell: The Extraordinary Communities That Arise in Disaster*. Many people act with compassion, generosity, and solidarity, others with fear, greed, and violence. Elites are particularly prone to act badly, Solnit argues, because they know that disasters reveal and shake up power dynamics. In contrast, the networks of mutual aid created by ordinary people "show us both what we want and what we can be." In such liminal moments, she says, "beliefs matter" because they shape how we act.[38]

George Lakoff spent his career studying people's beliefs. A professor of cognitive science and linguistics, he says that we make sense of the world through different "modes of thought," or worldviews. Those who gravitate to what he calls "the strict father model" are more likely to believe that the world is a dangerous, immoral place, hence the need for competition, discipline, and strong male leaders. After Hurricane Katrina, such people asserted that individual responsibility, such as evacuating before the storm, protected people better than government, which they call the nanny state. In contrast, those who tend toward the gender-neutral "nurturant parent model" value care, community, collaboration, and empathy. These people were outraged by the government's slow response after Katrina.[39]

Research shows that most people have aspects of both worldviews embedded in their psyches. How they see an issue such as climate change is influenced by the language, metaphors, and values used to frame it. Lakoff predicts that disappearing species, melting Arctic ice, and tragedies like Katrina will help humanity shift toward "a new consciousness," where instead of raiding the environment for profit we work to live within it safely.[40] He predicts that shared values will aid this shift more than statistics.

A 6'2" military man who has served in combat and on corporate boards, Lieutenant General Honoré is a master at speaking across the worldview divide. After Katrina, his blunt speaking style made him a media star along with the fact that he broke the logjam of inaction and finally got people evacuated. He was nicknamed the "Category Five General," the "Black John Wayne," and even the "Ragin' Cajun," although he identifies as "African American Creole"—from the mixed-race people of Louisiana whose ancestors were Black, white, and Indigenous. The General, as he prefers to be called, expresses deep compassion for anyone suffering from pollution regardless of race or political party. While he believes the fossil fuel industry needs government regulation and oversight, he speaks of American values in a way that resonates with the strict-father crowd.

"Now, don't you come down here talking about climate change!" he warned me on a phone call before my first visit in 2018. "Ask people where they fished when they were young. Then ask them what has happened to that place. Then you can talk about toxic pollution." In a softer voice he added, "People here have been programmed by Sean Hannity to think climate change is a China problem." He had found that pollution was easier for Louisianans of differing worldviews to see, smell, and feel first-hand.

My most generous guide in Louisiana, the General once drove me out to the Atchafalaya Basin, a vast swamp teeming with diverse species. We passed cypress trees, Confederate flags, and Trump signs on our long drive to a meeting of Cajuns concerned about the future of the crawfishing industry. They greeted the General warmly, and he gave a short, supportive speech to the full community center. Mostly we listened as fishermen shared how their livelihoods were threatened by oil industry laxity. Afterward, we were invited to a bustling home for a delicious crawfish boil, where I had to be taught how to dissect the local staple. I watched the General connect to their values of hard work and desire to protect their unique region. Instead of focusing on their differences, he appealed to their common identity as Louisianans.

THE HIDDEN COST OF BUSINESS

I thought of the General several times during the LMOGA conference. On my first visit to Louisiana, he gave me a "toxic tour" of Baton Rouge, which is both the state capital and a petrochemical hub. From a bluff on the campus of his alma mater, Southern University, we looked out over one of the Mississippi River's loping oxbow turns. There the river was two thousand feet wide and milk chocolate brown. "You see the trees over the horizon there?" asked the General, pointing to a pretty patch of green. "That land is polluted." He repeated those words several times as he drove me around his city. Even the lakes on the capitol grounds contained deadly polychlorinated biphenyls (PCBs), though the posted signs didn't keep people from fishing there.[41] The General acknowledged that much dumping occurred before people knew how dangerous such chemicals were. Now that they know, the companies should clean up "their junk," but they resisted taking responsibility because the pollution happened before the Clean Water Act. "What kind of shit is that? That's not right," he said.

Many of the General's stories were about companies trying to save money at the expense of public safety, as well as the turtles and fish, which the General included in his stories. He said industry's approach to pipeline maintenance was "like driving your car 'til your tires wear out." A pipeline rupture could do tremendous damage in a region with hundreds of waterways and tens of thousands of people employed in fishing. The General recalled pressing oil and gas leaders to improve pipeline safety, but they complained about their profit margins. "You didn't do it when you had $110 barrel of oil, man!"

THE TRUTH is, petrochemical companies would not have a profitable business model if they had to pay the full cost of their spills and leaks, if they had to pay the medical bills of those who get sick from living next to their facilities, and especially if they had to pay their fair portion of the $7.9 trillion dollars of damage climate change is expected to cause by mid-century. In economics, the costs foisted onto others are called "externalities." It's an interesting term. It suggests that damage to the environment, the community, or the climate is somehow external—separate from those making business decisions. But the consequences are not separate from them, or for that matter, those who work for them, live near their facilities, use their products, or buy their stocks.

Although I knew about externalities as a concept, their impact sunk in as the General drove me around the sprawling Exxon-

Mobil complex north of the capital. Built along the Mississippi by Standard Oil in 1909, the refinery is now buttressed by a series of plants, which turn petroleum by-products into ingredients used in diapers, chewing gum, tires, and makeup.[42] In addition to the pollution that is legally allowed, the complex has had hundreds of accidents and leaks.[43] In 2018 alone, ExxonMobil Baton Rouge released over 11.4 million tons of air pollution, including carcinogens such as benzene, not to mention climate warming gases.[44] Referring to one large accident where Exxon told community members to "shelter in place," a Louisiana Sierra Club organizer told me that locals call this advice "die where you at."

As we drove through the surrounding Black neighborhood, past abandoned buildings and empty lots, the General said, "This community paid the price for the wealth that multinationals have created and taken out of this state. They paid the price in terms of cancer and asthma. And many of them didn't get a chance to get those plant jobs." I asked if he thought industry could do more to reduce the health risks of their facilities, and he was adamant that they could. "But what they'll tell you is it dips into their profits." He added that executives get moved around. "They do not get connected to the community. They come in with one objective, to reduce cost and raise profits." I speculated that the executives don't live anywhere near their plants, and he agreed. They live in a more affluent neighborhood on the other side of town. That's also where the General lives, but he clearly felt connected to the people who lived closer to Exxon, whom he had met at church halls and other gatherings.

BETWEEN EXTERNALITIES and the cost of federal tax breaks, the International Monetary Fund concluded that US fossil fuel companies are subsidized to the tune of $700 billion per year.[45]

One of the things that most irked the General was that Louisiana remained one of the poorest states in the country despite being one of the highest producers of petrochemicals. "All the collusion with industry has *not* helped the economy of the state," he said. "Louisiana does the rest of the nation's dirty work." I noticed the General used the word *stupid* repeatedly when describing these issues. I asked if he thought the root problem isn't greed more than stupidity. "Oh, yeah. Hell, yeah. Absolutely. It does come down to greed."

Nothing I heard during the LMOGA conference convinced me otherwise. The only reference I heard to communities like the one near Exxon was during a panel on how to elect pro-industry judges to rule against what they saw as frivolous lawsuits. By the second day of the conference, I was ready for the activist disruption, if only to hear people articulate a worldview that valued people over profit. As an FBI agent spoke about cybersecurity, I nervously watched the ballroom doors, anticipating the confrontation.

INTERRUPTING THE ILLUSION

The ballroom doors burst open, and seven activists strode to the stage singing:

> *People gonna rise like the water. We gonna face this crisis now.*
> *I hear the voice of my great-granddaughter saying, "Keep it in the*
> *ground!"*[46]

There was confusion as everyone turned to see what was happening. Someone from the conference announced a break, and hundreds of suits stood up and headed for the exits. A few dozen men lingered near the doors to listen as the activists took turns making statements. The first was Anne Rolfes, who collaborated with frontline communities across Louisiana. Anne explained that LMOGA had lobbied for a new Louisiana law that made it a felony to protest on a pipeline construction site, punishable with fifteen years in prison. "So, we have come to your meeting to let you know that it's time for you, the oil industry, to stop destroying our state." I moved forward, heart quickening as I realized that taking photos made me far more conspicuous than my backpack and undyed hair. Nearby, an industry woman in a bright pink suit and long, styled hair filmed the protest as the New Orleans police arrived.

On average, the activists were younger than their intended audience, more androgynous in their dress, and the majority were female, with one member whose gender was ambiguous. This was a noticeable contrast to LMOGA, where the speakers and participants were both majority male. The activists' participatory speaking style reflected George Lakoff's nurturant parent worldview. Before being handcuffed, one said, "My husband and I are starting to plan our family. For your children and mine, we need to keep fossil fuels in the ground." As Anne was escorted out of the room, she spoke loudly about the fight against a new Formosa Plastics plant in Cancer Alley. Locals weren't even likely to get the jobs in that plant, she asserted as she was whisked away. "You have eleven years to avoid catastrophe," warned one millennial as police led her toward the doors. About fifteen minutes after the action began, the final two women were arrested, and conference participants were called to return to the ballroom.

As I moved back to my original table, I noticed many empty chairs and, more significantly, a palpable change in the energy, which no longer felt aloof from the outside world. The jovial arrogance had been punctured. So had my anonymity. The woman in the pink suit sat down next to me and drilled into me with her eyes as she introduced herself with a tight smile. Scribbling as she glanced at my name tag, she asked what I was doing at the conference. I explained I was writing a book and was interested in the perception gap between the fossil fuel industry and its critics. It was the truth, the reason I had paid $400 to attend the conference before I even knew there would be a protest. I also knew my answer did not capture the whole truth of my motivations. I was not an impartial observer.

"My last book was on climate change," I added, keeping my voice even. As soon as she googled me, she'd learn that I train people to do the kind of action we just witnessed. I suspected my neighbors at the round table had already done this, as a nearby man showed his phone screen to a woman whom I'd met at the conference happy hour the night before. From her pursed lips, I guessed they were looking at my website or maybe my profile on Twitter (later renamed X), which featured a banner I helped paint, with "Protect What You Love" in red and purple letters next to a cypress tree. The banner was from the campaign to stop the Bayou Bridge Pipeline, which was built to transport oil across seven hundred bodies of Louisiana water on its route from Texas to the Mississippi River.

A YEAR before the LMOGA conference, I visited Cherri Foytlin, who led the resistance to the pipeline. We sat outside as the sun lowered over the *L'Eau Est La Vie* camp, French for the Indigenous motto that anchored the struggle: *Water Is Life*. Cherri explained, "There's a handful of things that pull all people together, and water just happens to be one of them. We all come from water. Carried in our mother's womb in water. Every civilization since the dawn of time has gathered at water. Why? It's necessary for life." That is why they use the term "water protectors" instead of "activists."

Cherri, who describes herself as Dené, Cherokee, and Latina, hoped to build common ground between different groups threatened by the pipeline. This included the majority-Black community where the pipeline would end, landowners along the route, and Cajun crawfishermen who wanted to keep oil out of the water. Cherri said she felt for the Cajuns and recounted how the working-class French descendants got pushed out of Nova Scotia. Once they found a place to settle in the Atchafalaya Basin, they basically got sold off by France to become part of the United States. For the most part, Cajuns stayed out of the Civil War, she explained. "They thought it was a rich man's war, nine times out of ten. So fast forward to this time, and it's the same situation. They're stuck in the middle of a rich man's war. And unfortunately, the war is against them this time."

The war was also against Black communities like St. James Parish (a parish being the Louisiana equivalent of a county). In the heart of Cancer Alley, St. James already suffered from more than its fair share of petrochemical infrastructure, yet it was slated for a new Formosa Plastics plant in addition to the pipeline terminus. Just in the year and a half she'd been working to stop the pipeline, Cherri told me that she knew three people from St. James who had been diagnosed with cancer or passed away from it. Agencies just see a number, if they see anything at all, she said, but she sees the loss of people she knew and who knew her six kids. "I start to take it very personally when I see people get stepped on."

SOME ORGANIZATIONS pursued legal strategies to stop the pipeline. Meanwhile, Cherri purchased a piece of property in the pipeline's path to serve as a resistance camp, which received a blessing from the land's original stewards, the Atakapa-Ishak

Nation. The pipeline company, Energy Transfer Partners (ETP), rerouted around L'Eau Est La Vie camp, hoping to avoid a standoff like the one it faced at Standing Rock. Some landowners gave Cherri and other water protectors permission to nonviolently interrupt pipeline construction on their property. Over the months of construction and civil disobedience, water protectors were tased, pepper-sprayed, and choked by ETP security, which turned out to include off-duty cops as well as probation and parole officers. Some water protectors were brutally arrested and threatened with long jail terms through a new law that LMOGA had lobbied for.[47]

Despite courage and persistence, L'Eau Est La Vie never approached the size of Standing Rock. Still, the campaign punctured the illusion that all Louisianans support industry expansion. It also helped build connections that ultimately aided the successful campaign to stop the Formosa plant in St. James a few years later. By the time of the LMOGA conference, the pipeline was up and running. When I turned my attention away from the pink-suited lady and back to the ballroom stage, public relations specialist Jim Harris was talking about the very same pipeline.

THE ILLUSIONISTS RESPOND

Jim Harris was the guy companies hired when they had a public relations crisis. One of the few bearded men at LMOGA, he was a master of illusion. During the infamous BP spill, he deflected blame away from the negligent company and instead manufactured outrage at President Barack Obama for issuing a moratorium on deepwater drilling. Now, Harris was trying to regain control of the narrative after the protest. He acknowledged that the Bayou

Bridge Pipeline opponents had used effective messaging. I smiled faintly. To Cherri, *Water is Life* was wisdom or common sense, but for Harris, it was just "messaging." Of course, he didn't mention the brutal way Cherri was thrown to the ground during one of her arrests, or the fact that the pipeline company's private security intentionally capsized a boat of water protectors into water full of alligators and water moccasins. Instead, Harris named various organizations in the Louisiana environmental movement, emphasizing that "outsiders" had joined the pipeline fight.

Harris acknowledged but dismissed allegations of "environmental racism" by saying that many African Americans who live near petrochemical plants moved there after they were built. I thought of Mossville, which was founded in southwest Louisiana by an emancipated slave in 1790, long before the vibrant Black community was surrounded by over a dozen petrochemical plants. One journalist described Mossville as "quite possibly the most polluted corner of the most polluted region in one of the most polluted states in the country."[48] When people did move into such neighborhoods, it was usually all they could afford or find, given the realities of residential segregation and economic inequality. But Harris didn't mention any of the broader social conditions that led to the well-documented racial disparities in pollution exposure. Instead he claimed, "They don't have to tell the truth, and we do."

This was the most preposterous thing I heard at the two-day conference, making me wonder if Harris believed his own spin. Still, he was the first LMOGA speaker in two days to even acknowledge allegations of racial injustice or the growing movement to curtail fossil fuel use. While his doublespeak was exasperating, his comments were clearly in response to the activist speeches, not part of the presentation he had prepared. The nonviolent action had at least punctured their illusion of unassailability.

US Representative Garret Graves took the podium and openly admitted that his remarks were in response to the interruption. "We all live right here in this community, where your facilities and operations are," he said, though I suspected that no one in the room lived close enough to petrochemical facilities to smell the pungent odors or hear the emergency sirens. Expressing skepticism about the human causes of climate change, Graves acknowledged that increasing hurricanes and coastal erosion were happening, but he wanted to address them through "pro-growth incentive policies"—this meant without upsetting his donors in the oil and gas industry. "If we're making ourselves less safe by forcing the coast to erode, we're shooting ourselves in the foot," he continued. "I don't think any of us are stupid enough to shoot ourselves in the foot." Some old decisions were stupid in hindsight, but people didn't know better at the time, he claimed.

AS A whistleblower revealed, Exxon knew enough to start doing its own climate research in the 1970s. By the 1980s, the company knew that CO_2 levels were on track to double by 2060, which would dangerously increase global average temperatures.[49] Studies conducted for the company's internal use also affirmed that human activity was causing global warming. Yet, the company published advertorials—advertisements deceptively written to look like educational articles—that promoted a high degree of skepticism about climate change.[50] Their sophisticated efforts to confuse the public helped delay climate action. So did their funding of congressional candidates, especially powerful committee chairs. A scientist who read through and analyzed Exxon's internal documents noted, "With a strategy of delay, they could continue with a very profitable business."[51]

ExxonMobil Baton Rouge refinery head, Gloria Moncada, told the final LMOGA panel that if Exxon didn't get what it wanted in Louisiana, it would go elsewhere. "We have the ability to take a global view, being a global company. Geography isn't so defining anymore." This made my stomach turn. As indifferent as they were to the lives of people in the United States, the fossil fuel industry has an even deadlier track record in Africa, Asia, and Latin America. Moncada concluded that all their companies had a shared stake— presumably in rolling back regulation and avoiding taxes, two major concerns at the conference. "Going it alone doesn't work. We can move things if we work together," she said in the most disheartening appeal to common interest I had ever heard.

As the conference wrapped up, I grabbed my backpack and fled the Ritz-Carlton to meet the activists, who were celebrating their swift release at a noisy French Quarter bar a few blocks away. It was a relief to see them, not only because they'd been arrested, but because I'd felt so isolated at LMOGA. In contrast, the activists were glowing, exhilarated by taking action together, many for the first time. Anne introduced me, and they leaned forward to hear how their arrests had changed the conversation in the ballroom. One said she'd seen me taking photos and felt supported, even though she didn't know who I was. Over fried boudin balls, we exchanged stories and email addresses. They vowed to continue working together as I went to change out of my business dress and heels before enjoying some New Orleans jazz with a friend.

A few hours later, my stomach clenched when I noticed that the PR person for Exxon Baton Rouge had started following me on Twitter (now X), where my "Protect What You Love" banner proclaimed my worldview.

LESSONS FROM LMOGA

My father worked on a Sun Oil tanker in the 1950s, and two of my great-grandfathers were coal miners. I empathize with the need for a job to "put bread on the table," as the wife of an Exxon worker put it when I stayed with the couple in Baton Rouge. After visiting many people who were protesting fossil fuels, I wanted to learn more about those who worked for the industry. All of a sudden, opportunities started presenting themselves. A Shell worker, who moonlighted as a Lyft driver, told me he was thinking of quitting after a coworker fell into a deadly vat of chemicals, and the company tried to cover it up. I chose an Airbnb that catered to industry workers and heard more stories of men risking their lives to support their families. In each of these encounters, I was able to make a connection with the speaker, to feel empathy for their choices, even when we didn't share the same politics. In contrast, I felt painfully alienated from the LMOGA executives, who were risking other people's lives for their own profit.

At the conference happy hour, I tried approaching a few women but struggled to chitchat, even though I normally love engaging strangers. I believe that on a spiritual level, we are all part of a larger whole, but it felt like a thin fishline of connection, hard to see under the hotel chandeliers. Although I had gone to hear their side of the story, I couldn't stop the running critique in my head about how delusional *they were*. Did my Quaker belief that there was "that of God in every person" mean I had to see the divinity in *them*? I believed it did, but I wasn't sure how to do that, given that companies like Exxon were responsible for an unfathomable amount of harm. The human death toll from climate change could easily be in the hundreds of millions. I boiled down my dilemma to a challenging theological question: How do I love my neighbor when he is killing our other neighbors?

Back home I shared my LMOGA experience with the Quaker Support and Accountability Committee that meets with me regularly to help me stay in touch with Spirit's guidance for my work. One member said she could empathize with industry's feeling of victimhood, which I had pointed out with incredulity. "It is true that people want to sue them," she said. "It's true that people want them to pay more taxes," one of the recurring complaints at the conference. Put this way—by a woman of color who is quick to empathize with those hurt by industry—I could glimpse how executives felt beleaguered. From their worldview, they were just defending their profits, which was their obligation to their shareholders in a capitalist system.

As a thought experiment, I tried to find something I could see from their perspective. I remembered Jim Harris's claim that industry critics lied. In Louisiana, I'd heard Exxon described as a "$44 billion-a-year company," which was actually the company's net income before the 2008 recession, rounded up to the nearest billion. In 2018, they netted less than half that. For most people, any number in the billions sounds fantastical, but if you're Exxon management, you know these numbers vary wildly. Having fact-checked the wording of volunteer-written press releases, I knew many activists didn't pay attention to such details, but I could see how inaccuracy looked like lying to Jim Harris, the industry PR specialist. I could also see how my role at the conference looked duplicitous to the woman in the pink suit, even though I had been careful not to lie.

IT WAS humbling to acknowledge other ways I could relate to the LMOGA speakers. Not so long ago, I thought of the environment as a place to camp on vacation rather than the air I breathe and the water I drink. Not so long ago, I was unaware that my daughter's asthma was likely caused by an oil refinery in a neighborhood only a few miles away from our home. And not so long ago, I thought that choosing the "green" fund for my retirement account meant that I wasn't investing in fossil fuels, when in fact I was. It was easy to believe that I wasn't part of the problem, though, of course, that is part of the illusion of separation.

On the last morning of the LMOGA conference, I arrived late, and a server brought me a plate of bacon and eggs right before a break. I stuffed the bacon in my mouth and darted to the bathroom as quickly as I could in high heels, eager not to miss the impending arrival of the activists. With a mouth full of bacon, it occurred to me that I wasn't doing very well on my intention to cut down my meat consumption. Since industrial livestock production contributes to climate change, I resolved to try harder, even though I know that BP promoted the concept of individual "carbon footprints" to shift blame for climate change onto individuals.[52]

My husband and I make a conscious effort to reduce our fossil fuel use, but after many years of hanging our laundry to dry and other such actions, I've come to think of these efforts as spiritual practices meant to strengthen my mindfulness and integrity rather than as strategies for wider change. My change work focuses on challenging corporations, which have so much more power than individuals.

Through my own activism, I have seen how small groups of people can challenge powerful decision-makers through nonviolent strategies. But watching the effect of the activist disruption of LMOGA from inside the ballroom—both its bubble-bursting impact and the limitations of what it achieved—brought home the fact that such tactics, in isolation, are not enough to counter the inordinate power of a company like Exxon, which operates in two hundred countries. I came away from LMOGA convinced of something that Martin Luther King Jr. learned in the final years of his life: The change we need is both spiritual and systemic. Shifting our worldview to one of care and connection is necessary to build a just and sustainable society; so is learning to communicate across the worldview divide. But that's not all that's necessary. As Dr. King put it, we need both love and power.

CHAPTER TWO

POWER WITHOUT LOVE

The same week as the LMOGA conference, I visited the Vietnamese American Young Leaders Association of New Orleans (VAYLA), which was founded to build grassroots power after Hurricane Katrina. I came to the organization's storefront office to hear about their recent campaign against a new gas plant in their neighborhood. I sat on a folding chair across from Minh Nguyen, the executive director, and his colleague Mark Nguyen. They were both in their thirties.

There were different reasons people opposed the gas plant, they explained. Many argued that with two waterways, New Orleans East was too vulnerable to flooding. That would increase with climate change, to which the plant would contribute. Some emphasized that the construction cost would be borne by the city's mostly low-income customers, not Entergy, the profitable company that wanted the plant. A coalition formed, including local and national organizations. Together they turned out residents and supporters from across the city to testify at every hearing. They held marches. They filed lawsuits. Especially of interest to me, they built an intergenerational coalition that included Asian, Latino, Black, and white people, a level of diversity many coalitions struggle to achieve. "We know that we can't be anything by ourselves," said Minh.

MINH LEARNED this after Hurricane Katrina. While people across the region struggled to return to the neighborhoods they loved, the Vietnamese community was among the first to rebuild, strengthened by the knowledge that they had pooled their resources and started over once before when people like Minh's parents fled Vietnam in the mid-1970s.[53] As they worked to recover from the hurricane, the city of New Orleans decided to dump hazardous storm debris in their neighborhood, perilously close to a wildlife refuge and a canal that thousands of Vietnamese Americans drew on to water their vegetable gardens. Twenty-one at the time, Minh was one of the youths who galvanized intergenerational resistance to the dump. It was not part of Vietnamese culture to fight the government let alone have the old follow the young, but the elders joined, worried that if the dump ruined their tight-knit neighborhood, their children might move away forever.

After a protest, New Orleans Mayor Ray Nagin agreed to meet with them, but he quickly reneged on his promise to pause the dumping. At one point in the campaign, an experienced public interest attorney explained to Minh that the landfill was put there because the Vietnamese community had no power. They didn't even vote. In contrast, at least one of the New Orleans dumping companies had made a major campaign contribution to the mayor.[54]

The lesson about power stayed with Minh, who helped his team learn how to analyze the power dynamics of their situation. They recruited support from outside their community and participated in hearings, lawsuits, and sit-ins at the mayor's office. Finally, they blockaded the landfill itself. An old video clip shows Vietnamese women with pointed straw hats standing next to their Black neighbors shouting, "We are united, and we are powerful!"

They won, and the dumping was stopped.

MINH LEARNED from the struggle that his community had to start voting, but being relatively small, they also needed to work in coalition. "We have to come to some type of common understanding that we are all human," he told me. To this end, those opposing the gas plant did anti-racism training to help people find common ground despite the misperceptions they had of one another. Minh noted that many Black and Latino people thought of Asian Americans as the privileged minority, but Minh and Mark had both grown up in housing projects and on food stamps. Minh's parents were still fishermen, the skill they brought from Vietnam. "It took a lot of dialogue," he recalled. As they collaborated to stop the plant, they prayed together, laughed and cried together, and built lasting relationships. When there were conflicts, they dealt with them. "If we are fighting internally, we cannot fight outside. There's no way. So, we have got to figure out how to support each other and how to understand each other."

In the gas plant campaign, they were up against a multi-billion-dollar company that spent many millions "to fight little organizations like us," Minh said with a wry smile. Entergy gave out many small grants in the city, which bought the silence of community leaders of all races. One New Orleans City Council member had taken Entergy money to fund a parade. Another used to work for the utility and had consulted with Entergy before running for council.[55] Then there was the fact that one of the chief advisers to the city council was married to the deputy general counsel for Entergy.[56] The conflicts of interest were egregious. The city council voted six to one in favor of the gas plant.

After the plant was approved, the coalition discovered that an out-of-state Entergy subcontractor had paid scores of actors to arrive early to the hearings and fill all the seats. Mark was one of many residents locked out in the hall. Photos show a hearing room full of identical orange T-shirts. One printed poster read *Jobs, Giving, Community, Entergy.* The utility initially denied they were behind this, but one community member recognized an actor friend with the pro-industry group. Some actors broke their nondisclosure agreements to tell the truth. "They literally paid us under the table," one said of their payoff, which took place at a nearby Dave & Buster's. When the story broke, a local news outlet reported that the Entergy New Orleans president and CEO Charles Rice Jr. had chuckled at the room full of actors and said, "I think we've got them outnumbered."[57] After coverage of the controversy, Rice stepped down, but even that wasn't enough to reverse the plant's approval. Lawsuits were ultimately decided in favor of the company.[58]

POWER WITHOUT LOVE, AND LOVE WITHOUT POWER

I appreciated Minh and Mark's frank appraisal of their power—and its limits. Understanding how much power we have, and how to build more, is crucial to making change. Martin Luther King Jr. spoke about this on the tenth anniversary of his organization, the Southern Christian Leadership Conference. King acknowledged all they had achieved, including making racism an issue grappled with in every state capital. They had empowered Black people in the South to resist, a great achievement given the harsh repression of the Jim Crow era. They'd passed legislation, desegregated lunch counters, and won economic gains. Then King went on to acknowledge all that was left undone in the face of glaring economic inequality. He argued that what they needed was more power, which he described as "the strength required to bring about social, political, and economic change."

People of faith often eschewed power, King noted, having been told that power was incompatible with the love taught by religion. He argued that seeing power and love as opposites was based on fuzzy philosophy. "Now, we got to get this thing right," he proclaimed. "What is needed is a realization that power without love is reckless and abusive, and that love without power is sentimental and anemic." Shouts of yes and applause accompanied him as his voice rose. "Power at its best is love implementing the demands of justice, and justice at its best is love correcting everything that stands against love. And this is what we must see as we move on."[59]

King's assessment of the civil rights struggle in 1967—a year before his assassination—has important lessons for us today. Certainly, the fossil fuel industry exemplifies the truth that "power without love is reckless and abusive." And it's not unfair to say that the early climate movement was "sentimental and anemic," naively hoping that politicians would be swayed by scientific projections and images of our blue-and-white planet taken from space. In recent years, the climate movement has made significant strides, including concrete wins on concrete issues. Still, oil, coal, and gas continue to be wrenched from the Earth. Infrastructure continues to be built to process, burn, and transport these fuels, even though scientists have told us we must transition away from them. During my travels, I met many groups like VAYLA New Orleans, full of commitment, skill, and love, but lacking enough power to thwart a multibillion-dollar corporation.

IDEALLY, ONE role of government is to protect the common good, but Mark mentioned that many people who were "on the other side" on climate change were in positions of power. Although our conversation occurred during Donald Trump's first presidency—when Exxon's former CEO was appointed as secretary of state—the revolving door between government and corporations is a bipartisan problem. After all, the Entergy gas plant was supported by Democrats in the New Orleans City Council.

Many people who work for environmental regulatory agencies are sincere advocates, trying to do their best in a dysfunctional system. Some are former industry people with divided loyalties, like Chuck Carr Brown, Louisiana's top environmental regulator, who supported the Entergy gas plant and promised LMOGA participants that he wouldn't fine them much. EPA Regional Administrator Anne Idsal was even more solicitous at LMOGA. She invited industry leaders to offer her feedback on "how we provide customer service to y'all." During her years at the EPA, Idsal—whose name became Austin after she married—would help roll back methane regulation on the oil and gas industry and support other policies that made it easier for fossil fuel companies to emit carbon dioxide.[60]

During the conference, Idsal said that people in industry had told her that in the past they didn't feel they had a seat at the table. They felt they were "on the menu." The audience chuckled. I thought of Eddie Bautista, who used the phrase "If you're not at the table, you're probably on the menu" to explain why survivors of Hurricane Sandy wanted to influence plans to rebuild New York. I'd also heard the phrase from those dealing with the high rates of cancer and heart disease that cluster around polluting facilities. Being at the table meant having a meaningful voice in decisions that were life or death for such communities. This was not the same as having token racial representation, which often does not translate into collective political power. For example, Chuck Carr Brown and Entergy New Orleans CEO Charles Rice Jr. were both Black, but gas plant opponents in New Orleans East did not feel those men represented their interests.

The fossil fuel industry uses campaign contributions and the promise of jobs to buy their place at the tables of power. Yet the LMOGA audience chuckled at Idsal's implication that they themselves had been "on the menu." For industry, it was an eat-or-be-eaten world, and any attempt to challenge their impunity was seen as an attack. One of the most alarming LMOGA sessions I attended was about how to get pro-industry judges elected "to put the final nail in the trial lawyers' coffins," as one panelist put it. Over the coming years, as Donald Trump's second administration attacked the legal framework for environmental protection, the pillars of power theory would help me understand more clearly why industry wanted to have judges in their pockets.

WEAKENING THE PILLARS OF POWER

The patterns I'd heard at both VAYLA and LMOGA fit perfectly into a theory I often teach. It's based on research that shows that even dictators need people's cooperation to maintain control. Quaker activist and trainer George Lakey took this insight a step further, arguing that repressive systems rely on the cooperation of institutions as well as individuals. One strategy for making change, especially when the power holder seems invincible, is to weaken the support they receive from what George called the pillars of power.

When sharing this theory in front of a dry-erase board, trainers often draw a pyramid to represent how we usually think of power—wide with a stable bottom that looks impossible to move and a narrow top that's hard to reach. Then we draw an upside-down pyramid, which needs pillars to keep it stable, which is how power actually works.

Our image of power

How power really works

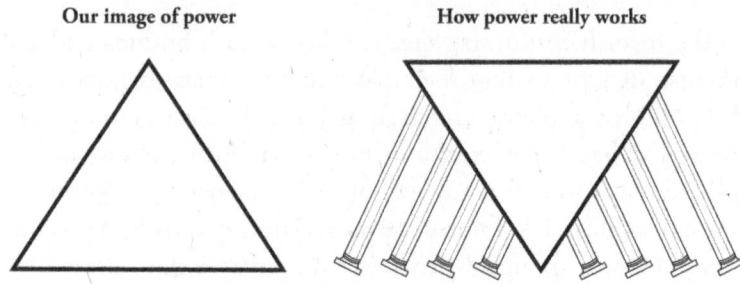

I often ask groups to brainstorm some institutions that buttress the fossil fuel industry's power. They usually name banks, which finance industry projects, as well as pro-industry politicians, utilities, judges, and media. While the list of pillars is long, the model offers a strategy for change. When people successfully pressure those institutions to remove their support, the powerful become vulnerable. In the best-case scenario, different groups take on different pillars, as they did in the struggle for democracy in South Africa.

Many know that the word *apartheid* means "separateness," which was foundational to South Africa's racist regime. For nearly five decades, apartheid was propped up by draconian laws, violence, a tightly controlled media, foreign aid, and the complicity of international companies operating in the country. Black women organized against the pass system that separated their families. Actors violated racial segregation laws by performing to integrated audiences. Students boycotted the segregated school system, and workers unionized to build collective power. Many South Africans were killed for resisting. Outside the country, allies organized sport boycotts and sanctions to undermine apartheid's legitimacy. Students in the United States pressured universities to divest their endowments from companies doing business in South

Africa. When resistance leader Nelson Mandela walked out of prison in 1990 after twenty-seven years, it looked like an overnight miracle to those just tuning in to the story.

Some people emphasize that Mandela, known affectionately as Madiba, took the courageous step of reaching out to his captors to negotiate his release. After he became South Africa's first Black president in 1994, he built bridges with whites, even inviting to tea the widows of the prime ministers who imprisoned him. Those are inspiring examples of leadership that put reconciliation over retribution, but it's important to understand that Mandela could only make those overtures because ordinary people had weakened the pillars that held up apartheid. Like Dr. King, Madiba recognized that both power and love were needed to build a just society.

ANTI-APARTHEID protests were among the first I attended in my twenties, when I was in grad school. Since my university refused to fully divest, I assumed the divestment movement was not effective. Much later, I realized that the fact that American students were willing to risk their professional futures in solidarity with people on the other side of the world did build pressure. Within a few years, over 150 higher education institutions divested, in full or in part. This weakened financial and political support for apartheid, contributing to its fall.

The pillars of power framework also clarifies what the coalition against the New Orleans gas plant accomplished. For starters, they effectively revealed the unfairness of the system. At one well-attended city council session on the gas plant, Rev. Gregory Manning testified that a council member had privately told him that the decision would not be influenced by public comments. "I submit to you today that it is immoral to ask grandparents, grandmothers and grandfathers, to take up seven hours of their day to come down here and for you to say that their comments really don't matter," said Manning. A Black pastor in a white clerical collar, his voice rose with the refrain that the council's actions had repeatedly been immoral. He was told his time was up, but people in the audience ceded their time to him, a sign to me that people were putting a strong collective narrative ahead of individual egos. Manning declared that it was immoral for any council member to cast a vote on the gas plant if they had ever received money from Entergy.[61]

The illusionists of industry rely on our belief that the system works fairly. Shining a light on its corruption can make politicians more vulnerable come election time. It can also put a wedge between industry and their political allies, which happened after the Entergy actor scandal. City council fined the company $5 million, and the council president said publicly that it would take a long time for Entergy to regain the public's trust, if it ever did. Better than anything the activists could have said, the fact that the company had to pay people to pretend to support the gas plant revealed how few people actually did. The revelation came too late to stop the plant's approval, but it did weaken a pillar of industry power.

After the plant was built, Hurricane Ida took out power in most of New Orleans. When some people suffered without electricity for a week, *The New York Times* pointed out that the new gas plant had failed to deliver on Entergy's promises. Recounting the various controversies around the plant's approval, the article quoted the president of a neighborhood organization, who said, "We, the citizens and the ratepayers that were against the plant, were correct."[62]

Feeling vindicated is small comfort in such a situation, unless we see it as part of building long-term power. Framed that way, vindication can strengthen the grassroots case the next time there is a proposed project. The gas plant coalition also raised the cost of doing business, something companies consider when deciding whether to pursue such projects in the future.

BUILDING POWER OVER TIME

Minh explained that, despite the disappointing loss, he felt it was important for people in his community to recognize their own power so they would continue taking action on the issues affecting them. That's why VAYLA translated into Vietnamese a list of all the ways the gas plant campaign had made a difference. "Entergy had to hire actors because of our organizing efforts," he stressed. Because they revealed the actor scandal, Entergy's CEO was demoted, and the company was fined $5 million. Some of that money would go to New Orleans East. The coalition also won commitments to renewable energy and energy efficiency. "There will be people coming in to check the emissions, the air quality, the pollution," Mark noted. The results would be shared with the community. Although they had lost the battle to stop the plant, they had built skills, relationships, and grassroots power, power they intended to use in future campaigns. That was important to acknowledge and celebrate, Minh said.

THE TRUTH is, we never know what our long-term impact will be. Consider the Baton Rouge bus boycott of 1953. Most people haven't heard of this early civil rights campaign because it ended after a week with a compromise that gave Black people access to more bus seats but didn't end segregation on the buses. Two years later, when Rosa Parks's arrest sparked a bus boycott in Montgomery, Alabama, people from Baton Rouge shared their realization that people couldn't sustain a bus boycott without another way to get to work. By arranging rides for Black workers, the Montgomery campaign was able to keep the boycott up for over a year, ultimately winning desegregation of the buses.[63] Even more significant, by empowering Black people to fight segregation and win, the Montgomery bus boycott inspired more nonviolent campaigns. It also launched the activist leadership of Martin Luther King Jr., a young pastor who was recruited to be the boycott's spokesperson.

The freedom movement, as the civil rights movement was originally called, has much to teach us today about how targeted campaigns against specific pillars of power can build broad momentum for change. Although racial segregation was the norm across institutions in the Jim Crow South, people in Montgomery started by focusing on the buses, where customers had some economic leverage through collective boycotting. Other successful efforts focused on school desegregation and voting rights, undermining important pillars of Jim Crow's power. Many people today assume that voting is the only power they have, but voting was restricted for many Black people at the beginning of the freedom movement, so they needed to find other ways to build power while fighting for the right to vote.

A major strategy was to challenge injustice directly and nonviolently by refusing to cooperate with it. It was illegal for a Black person to sit at a Southern Woolworth's lunch counter and order a cup of coffee, so that was exactly what four college students did in Greensboro, North Carolina. Soon lunch counter sits-ins galvanized students across the South, as well as allies who picketed in the North. Woolworth's announced a new desegregation policy several months later. Black and white Freedom Riders broke segregation laws by riding on buses together, even in the face of violent attacks. Black bathers waded into the ocean in defiance of whites-only beach signs while fully clothed police officers chased them into the waves to arrest them. Each of these tactics highlighted a specific aspect of segregation in a way that revealed the whole system's brutality, and sometimes its ridiculousness.

Such tactics are examples of nonviolent direct action. Whether or not it includes breaking the law, this approach breaks out of the deferential ways we are trained to behave in the face of power. When applied strategically and persistently, nonviolent direct action can generate enough public pressure to push decision-makers to make changes they otherwise would not. In addition to creating pressure, nonviolent direct action can move hearts and minds by dramatizing what is wrong and building sympathy for those calling for change. I believe this is especially true when action is grounded in the kind of love Dr. King spoke about in his 1967 speech. Such "strong, demanding love" was grounded in a vision of Beloved Community that included one's opponents. That is why King said his followers should attack forces of evil, not individuals doing evil. I found this teaching helpful when reflecting on my alienation at the LMOGA conference.

OF COURSE, changing a pervasive system one institution at a time is slow and hard. That is where mass mobilization can help. Eight years after the beginning of the Montgomery bus boycott, several organizations came together to organize the 1963 March on Washington for Jobs and Freedom. Despite skepticism when organizer Bayard Rustin predicted a crowd of one hundred thousand, an unprecedented 250,000 people came to DC, taking trains, planes, cars, and buses from across the country. A majority were Black people, who had a new sense of their collective power as they filled the grass between the Lincoln and Washington Memorials. Nine months after the historic speeches and music,

Congress passed the landmark Civil Rights Act, which prohibits discrimination based on race, color, sex, religion, or national origin. A year later, after a historic march from Selma to Montgomery, Congress passed the Voting Rights Act. As King himself noted, the laws did not end racism or the entrenched economic inequality between the races. Still, they were important changes that would not have happened without people power.

My generation was inspired by the 1963 March on Washington but failed to understand that it wasn't the one-day event by itself that was powerful. It was the ongoing and courageous campaigns to undermine the pillars upholding segregation, strengthened by a mobilization that demonstrated the movement's size and breadth. I didn't understand this history when I dragged my young children to freezing cold marches against the invasion of Iraq, only to feel discouraged when the invasion happened anyway. If we had all gone home and blocked Army recruiting offices, or refused to pay taxes, we might have had a better chance of preventing that deadly and expensive war. Instead, I assumed I was unable to impact the issues I cared about, including climate change. Then, I found a group that used the pillars of power theory to maximize a small group's impact, understanding that it was just a part of a larger whole.

THE MONEY PILLAR

Earth Quaker Action Team (EQAT, pronounced *equate*) was founded in 2010 after a regional gathering of Quakers the previous summer. The evening speakers shared different approaches to addressing climate change and environmental destruction, including personal conservation and lobbying. One speaker,

George Lakey, had spent his life training people around the world in the kind of nonviolent direct action he'd learned as a young white man in the civil rights movement. "If we're really serious that climate change is going to be catastrophic, shouldn't we try the kind of strategies that have worked historically?" George asked, gesturing with his long arms for emphasis. He gave examples of Quakers through history who put their bodies in the way of injustice, such as those who sailed across the Pacific in the 1950s to physically interrupt nuclear testing. Their act had an impact out of proportion to their size, helping to get nuclear testing banned. I was one of many in the large hall who felt deeply moved, sensing we were being called by Spirit to act more courageously.

A small group started meeting after George's talk. They used the pillars of power theory to identify potential institutions to pressure. One important pillar holding up the fossil fuel industry is money, with large amounts of capital needed for extraction projects. EQAT learned that despite calling itself a "green bank," PNC was one of the top financers of companies engaged in mountaintop removal coal mining, a practice just as reckless and abusive as it sounds. Many Quaker institutions kept their funds in the Pennsylvania-based bank, which bragged about its Quaker roots. Three EQAT leaders sat down with local bank executives and asked that they stop financing the destruction of Appalachian mountains and communities, which also contributed to global warming. When the bankers said they never made decisions based on anything but money, EQAT started pressuring PNC publicly, calling the campaign Bank Like Appalachia Matters! (BLAM!).

I didn't get involved until 2011, when a strong intuition led me to attend the Philadelphia Flower Show, which was sponsored by PNC that year. I was straining to see the garden displays in the crowded convention center when I ran into a Quaker friend who told me to come to the PNC display at noon. "Pray for us, and leave if the police tell you to," he warned before disappearing into the crowd. At noon, I found several people joyously singing "Where Have All the Flowers Gone?" Their matching T-shirts and yellow crime scene tape attracted curious bystanders and the media.

While it was climate change that initially motivated my involvement, Appalachians had long defended their mountains and streams from the more local and immediate effects of extreme coal extraction. They testified at hearings about the high rates of cancer and birth defects in their communities, as well as the disrupted cemeteries and ecosystems destroyed by coal company greed. They marched and were arrested, facing brutal treatment in jail and even death threats. Many of us in Philadelphia had no idea before joining BLAM! that EQAT members joined a West Virginia march linking the historic labor struggle on Blair Mountain to the contemporary struggle to stop mountaintop removal coal mining. Another time, two West Virginia activists offered a Pennsylvania PNC bank manager a jar of brown polluted water they had carried from home. It was a poignant way to dramatize the connection between PNC's investments, its customers, and the people of Appalachia. Many Quaker individuals and institutions started moving their money out of the bank. So did the Pennsylvania Sierra Club.

Using nonviolent direct action meant that we went right into places where PNC did business, holding up a photo of barren, razed mountains, or praying in silence for PNC to do better. Once we sang the old union song "Which Side Are You On?" to the PNC board during its annual shareholder meeting. We were challenging the hidden assumption of the banking industry—that profitability is the only criteria for investment, and customers should not question what a bank does with their money. Periodically, like during the shareholder and Flower Show actions, we defied police orders to leave. Risking arrest in this way conveyed just how morally reprehensible mountaintop removal was and pushed bank leaders to grapple with an issue they had felt separate from.

After two years, we felt the need to increase the pressure, so we discussed visiting PNC executives and board members at their homes, or at public events where they were speaking. This kind of tactic is usually called "bird-dogging," but some Quakers were uncomfortable with a term they thought connoted hunting. After some discussion, we decided to adapt this tactic, renaming it "spotlighting" to convey that we were not attacking individuals but making visible the consequences of PNC's loans. On one occasion, PNC board member Jane Pepper was leading a garden tour of England, and we tried to send her off from the Philadelphia airport with a large banner with a photo of decapitated mountains and smaller signs urging her to "Be Brave" and speak up. Although we missed her in Philadelphia, we recruited four English Quakers to show up at Pepper's hotel with a gift basket of chocolates and information on the dangers of mountaintop removal coal mining. Their friendly but earnest attempt to speak with her showed that she couldn't escape her connection to the policies of the bank on whose board she served.

In late 2013, I heard that Quaker teenagers in Florida wanted to connect to the BLAM! campaign. Then board clerk (or chair), I began an email exchange, unsure where it would lead. A few months later, we learned that PNC was moving its 2014 shareholder meeting from its Pittsburgh headquarters to Tampa, Florida. Quakers call this kind of serendipity "way opening." EQAT sent a few leaders to Tampa to support an intergenerational group to hold an action outside. Although I was the only person they allowed into the shareholder meeting, where I stood up and prayed in silence, PNC leaders rushed through their agenda in fifteen minutes, clearly worried about our presence.

A few months later, a national Quaker conference happened to be near Pittsburgh, another dramatic example of way opening. The Florida teenagers helped recruit two hundred Quakers to come to PNC's downtown headquarters. Doing follow-up trainings around the country, we were able to pull off a day of action in thirteen states and the District of Columbia in December 2014. Called "Flood PNC," it included thirty-one PNC actions on one day, with the Florida teenagers playing key action roles.

In March 2015, PNC adopted a policy moving away from its financing of mountaintop removal coal mining. The bank acknowledged health and environmental concerns, as well as financial ones, even though they had told us that they never made investment decisions based on anything other than money. At the time, we didn't even have a full-time staff person, let alone an office, but we moved a bank that netted over $4 billion that year. Later we heard from someone with inside information that our combination of boldness and spiritual grounding made us hard to dismiss.

UNDER PRESSURE from other groups, JPMorgan Chase, Wells Fargo, and Bank of America also agreed to stop financing major mountaintop removal companies. With the industry already financially vulnerable, two of these coal companies declared bankruptcy a few months after PNC's announcement. Although weakening the financial pillar didn't end the reckless and abusive practice of mountaintop removal (which later got a boost from the first Trump administration through regulatory rollbacks), coal production from Appalachia has been in decline overall.[64]

Targeting financial institutions has become much more common as more people have realized that money is a key pillar supporting the fossil fuel industry, especially since new fossil fuel projects require large amounts of capital. While companies like Entergy rely on amenable government officials to get permits approved, banks need to win the trust of many kinds of customers to be successful. That means their public image is extremely valuable to them, giving activists leverage. The same is true of investment companies, like Vanguard or State Street. As in the anti-apartheid movement, many groups today are pressuring their institutions to divest from fossil fuels. So far, over 1,600 institutions have committed. Their collective value is estimated at a whopping $40.76 trillion.[65]

Skeptics point out that divestment does not keep oil, coal, or gas in the ground, which is true, but they are missing the way this strategy works. When nuns, university administrators, and major cities declare that they can't morally invest in fossil fuels, it makes it just a bit harder for politicians and judges to take industry's money and do its bidding. When fewer banks will loan money for new fossil fuel projects, it raises the cost of such projects, giving more leverage to those, like VAYLA, who are fighting to keep them out of their neighborhoods. Divestment is a long-term strategy that works to strengthen other parts of the movement.

In recent years, there has been a backlash against sustainable investing, orchestrated by what Morningstar describes as "conservative activists who oppose things like climate action; diversity, equity, and inclusion policies; better worker pay and benefits" and other issues they consider liberal. As a result of pressure from the opposite direction, PNC and other banks have been rolling back their environmental and climate commitments, which is a sobering reminder that effective activism can result in pushback.[66]

THE ROLE OF RISK

While there has been a swell of nonviolent direct action in recent years, the threat of state violence and intimidation discourages many groups from using it, including the Vietnamese community in New Orleans East. Even though blocking the dump driveway worked for them after Hurricane Katrina, Minh and Mark knew that type of tactic was a hard sell for their base. They were a community founded by refugees for whom challenging authority was still countercultural. Although they hadn't ruled out nonviolent direct action, Minh and Mark told me that they were focused on building power through coalition and relationship building.

Their concerns about nonviolent direct action are understandable. There are many examples of law enforcement responding more aggressively to protests led by people of color. Rev. Manning was unexpectedly arrested a year after his testimony against the gas plant. At the end of a two-week march to highlight high death rates along the petrochemical corridor, a diverse crowd gathered with signs in the hallway outside the Louisiana Association of Business and Industry. An older Black woman was sharing that seven of her family members had died of cancer just that year when Baton Rouge police arrived and asked the group to leave. Rev. Manning, the emcee, suggested that the woman be allowed to finish speaking. When the crowd shouted, "Let her speak," the police moved toward Manning, then twisted his arm behind his back. The legally blind pastor felt his legs pushed out from under him. He was tightly handcuffed face down on the carpet.

In addition to two misdemeanors, Manning was charged with "inciting a riot"—a felony—though prosecutors later refused the charges. Afterward, Manning joked that if he was inciting a riot, it was what he did every Sunday morning. Although he was part of many environmental justice protests, this was his first arrest and it was unexpected. He learned that he and other activists needed more training in nonviolence to prepare for different scenarios. That said, he felt being arrested for "a righteous act" had a positive impact in bringing attention to the people who were dying because of pollution. "While I would have rather not been in an orange jump suit and strip-searched, and been in jail for eleven hours, it was somehow necessary to bring momentum to the movement."[67]

This is one of the core principles of Kingian nonviolence, an approach based on Dr. King's teachings. In *Healing Resistance: A Radically Different Response to Harm*, Kazu Haga writes that the Kingian principle that gets the most pushback in his workshops is "Accept suffering without retaliation for the sake of the cause to achieve the goal." People assume it means passively suffering, or saying that oppressed people should suffer even more. Instead, Haga shares what he learned from leaders of the civil rights movement, that Black people in the Jim Crow South were already suffering, were already in danger of violence just for existing. By choosing to take the risk of publicly standing up for justice, "civil rights activists were able to use that suffering to push the movement forward." For example, in 1965, peaceful marchers defied orders not to cross the bridge out of Selma. Alabama state troopers attacked the marchers, who did not fight back. By exposing the brutality of the state, Bloody Sunday "woke up the conscience of a nation," writes Haga. "Two days later, 2,500 people descended on Selma in support of the movement."[68]

IN MY work as a trainer, I've seen white people talk cavalierly about the possibility of civil disobedience, without recognizing that it may be much scarier for people whose communities have been targeted by police violence. Since the Black Lives Matter movement raised awareness of this issue, I've also seen people make the opposite mistake, suggesting that Black or Brown people "can't" do nonviolent direct action. I've even heard "We shouldn't let them" take that risk. This is patronizing and discounts the courage and agency of people like Rev. Manning. I find it helpful to acknowledge that risk is not equal, and then ask people to discern for themselves what risk they feel called to take. In addition to people of color, those who don't have citizenship or who have serious health conditions face greater danger when arrested. Jails usually separate people by gender, creating special risks for people who are trans or don't conform to the gender binary.

There are many factors that influence police behavior, including how much training they have and whether they feel threatened by a particular protest. When faced with civil disobedience against a corporation, they often take their cues from the company. In my experience, banks and those that prize their public image generally don't want the bad publicity that violent arrests would bring, whereas pipeline companies care more about their construction schedules and can be brutal toward those who try to stand in their way. When such things happen, activists can undermine their power by showing that law enforcement is supporting corporate interests over the public interest.

Taking a risk for one's beliefs can build support among those who hear about it. In Britain, after thousands of climate activists were arrested for civil disobedience, one judge speculated that judges were being rotated because they could only listen to so many of the activist testimonies "before our sympathies start to overwhelm us."[69] Other nonviolent campaigners have heard similar responses from public officials and even from workers at the companies they are protesting.

HOW SEPARATION HINDERS US

EQAT was founded by Philadelphia Quakers, a community that is majority, but not entirely, white and middle-class. For most of its history, EQAT has included few people of color, some whites with significant cross-racial experience and many with relatively little. Although we chose to target PNC because we saw a strategic opportunity to pressure a local, Quaker-founded bank, some Quakers of African descent criticized our focus on majority-white Appalachia when there were environmental justice issues in our own majority-Black city. I understood their perspective. Even within the campaign we chose, EQAT could have done more to connect mountaintop removal to the struggles of people in our region. In Philadelphia as elsewhere, everyone is affected by rising temperatures, though predominantly Black neighborhoods are disproportionately impacted by heat deaths, since those neighborhoods are less likely to have cooling trees and more likely to include people who can't afford air-conditioning.[70]

More broadly, we could have highlighted the coal industry's long history of pitting Black and white people against each other to undermine union organizing, especially in Appalachia. In recent decades, Appalachians have been poisoned by the heavy metals released into waterways by coal extraction, but the plants that burn coal were often built in Black and Brown communities in other regions. Same with the dumps that house leftover toxic coal ash. One of the things I learned from the BLAM! campaign was how interconnected these systems are, with climate change motivating more people to connect the dots.

As we approached the end of the campaign, we started to imagine what we might do next. Two themes emerged. First, we wanted to explicitly promote racial justice. Second, we wanted to take on a solutions campaign, one that not only protested what was bad but that encouraged the more just and sustainable alternatives. When PNC announced its move away from financing mountaintop removal coal mining in March 2015, we began a process to discern how to combine these priorities in a campaign.

Inspired by solar-jobs programs in communities like Oakland, we focused on green-job creation. The Black pastors in Philadelphia whom we met with for feedback affirmed the need for good jobs in their neighborhoods and shared how owning solar on their church roofs would enable them to save on electricity, so they could put more money into youth programming and other needs. One Black Quaker friend was also encouraging solar as a path to economic development for her North Philadelphia neighborhood. As a nonviolent direct action group, we knew EQAT's role would be applying pressure to power holders, so we used the pillars of power theory to identify our local electric utility, PECO, as a key player in deciding where our region's energy comes from. As other ideas fell away, we decided to push PECO to prioritize local solar both to slow climate change and to create economic opportunity in high-unemployment neighborhoods.

Bishop Dwayne Royster was excited that our Power Local Green Jobs campaign connected climate change and local justice, so the organization he led joined as a partner—Philadelphians Organized to Witness, Empower, and Rebuild (POWER). This offer of partnership from an interracial and interfaith group felt like another experience of way opening. I thought of our collaboration as "the crisis of the Earth saving us from our illusion of separation," coming about a year after I first heard that spiritual message.

With support from POWER, EQAT showed up to PECO headquarters relentlessly. We walked a hundred-mile loop around the utility's service territory, organized a few civil disobedience actions, challenged executives both in private meetings and in public. I found that my experience at LMOGA helped me to acknowledge the pressure corporate leaders felt to maximize profit, even as I urged them make decisions that would be better for their own children, as well as the communities where they operated.

As a result of our pressure, PECO leaders took several steps. They created a new solar department, donated at least $200,000 to solar job training, began to upgrade the grid to accommodate solar, and committed to buy more solar from local sources. Eventually we realized that they were not going to budge on our ambitious goal of 20 percent electricity from local solar by 2025. When we decided to refocus our energy after six years and ninety-three nonviolent actions, one solar innovator commented that before our campaign PECO was twenty years behind other utilities, and now they were only ten years behind. We had built enough power to move them, but not as much as we'd hoped.

Like VAYLA, we tried to celebrate what we had accomplished while frankly analyzing why we were unable to get a larger energy transformation. The utility was supported by many pillars, including regulations that protected their interests. The fact that PECO was owned by a parent corporation also muddied our claims that they had the power to do what we wanted. Showing they did have some agency, PECO drafted a low-income solar pilot project, but dropped the idea after the first Trump administration canceled Obama administration funding. Although they netted over a million dollars a day in profit, their leaders claimed that they could not use any of that money to invest in solar. Even the patchwork system of the electrical grid weakened our leverage. Each of these forms of separation gave the company excuses that undercut our narrative that they needed to do much more.

We couldn't blame everything on the system. We struggled to stay creative and bold, especially after their security started locking us out of their headquarters. We also never attracted the large or diverse numbers we had originally hoped for. Although we developed a good relationship with POWER leadership, we didn't inspire broad participation from their base of over sixty congregations. We also didn't mobilize as many Quakers as we did in our first campaign, even locally. In hindsight, our overly wonky demands didn't help. Perhaps our goals would have had more resonance if they had emerged not from the imagination of EQATers, but from a diverse grassroots process like the one Eddie Bautista shepherded in New York City after Hurricane Sandy. The people who most needed green jobs were not at the table when the Power Local Green Jobs campaign was created.

When the campaign ended, one EQAT member observed that we lacked the fervor of early AIDS activists, who knew they were fighting for their own lives and for those they loved. For many older EQAT members, the love that drove them was for their grandchildren, threatened by climate catastrophe, whereas their concern for jobs in neighborhoods like North Philadelphia was based on principle, not proximity. Working on the campaign deepened their awareness of racial inequality, but it also exposed the painful distance many of us felt from neighborhoods only a few miles from our own. It was yet another example of the illusion of separation.

LESSONS FROM VAYLA

I first met VAYLA leaders several months before my interview with Minh and Mark. When friends at POWER were invited to speak on a panel about environmental justice at a Sierra Club conference in New Orleans, they asked if I would go to represent the campaign. It felt like way opening when the panel organizer, Leslie Fields, spontaneously invited me to join a side trip to New Orleans East. At the time, I had only heard about the Vietnamese community there, and was grateful to visit with people from Kenya, India, and Vietnam as well as a few Sierra Club staff, including Leslie, an attorney with pulled-back dreadlocks. It was an unusually cold fall day as we toured the vibrant rows of vegetables at the co-op farm, one of the projects they started after Katrina. "It was the Wild West," Leslie said, recalling the post-Katrina period when dumping hurricane debris became a rogue business. It was the fight against the dump that brought together the Vietnamese community and the Sierra Club.

After the co-op farm, we headed to a nearby Vietnamese restaurant. At a long table, we leaned over bowls of pho as Minh recounted his community's history of resistance. Minh said that the lessons from the Entergy campaign would be useful in a struggle developing upriver in St. James Parish, in the heart of Cancer Alley. When he mentioned that people were mobilizing to oppose an enormous plastics plant that Formosa wanted to build there, Gwen, a quiet young activist from Vietnam became animated. She told us that Formosa had released toxic waste off the coast of Vietnam two years earlier, killing at least seventy tons of fish. She described masses of white bellies floating on the water and washing up on shore. The tragedy devastated tens of thousands of people who relied on fishing for income and their own food. Gwen had been part of protests that forced the company to reluctantly admit its responsibility.

Minh asked how they could connect what was happening in Louisiana with what was happening in Vietnam and around the world. Joe Athialy, the executive director of an India-based organization that monitors financial institutions, said that many people around the world were in similar struggles, sometimes even against the same companies. They had to find ways to share what worked and what didn't, so they could learn from each other. "This is everyone's struggle," he concluded.

"We need you," Minh said to Joe. Everyone nodded.

"We need each other," I added.

It's one of the most important lessons the crisis of the Earth has to teach.

WE NEED each other, but we often don't realize it because of the ways we've been divided. Class segregation not only keeps fossil fuel executives and members of Congress from smelling the chemicals released from their plants, it also hides the truth from the rest of us, those neither most responsible nor most impacted. This hinders broad coalition building. Racial inequality makes it more difficult to build power through nonviolent direct action because most people of color don't want to risk being treated like Rev. Manning, and most white people don't yet experience these issues as life or death. Ideological divisions make other differences even harder to bridge. I began to realize that these forms of separation are themselves pillars, holding up the petrochemical industry and weakening grassroots organizations like VAYLA and EQAT, though in different ways.

In subsequent years, I watched separation play an increasingly destructive role in American politics. After Donald Trump's 2025 inauguration, I shared the pillars of power theory in trainings and speeches to help people understand how he was attempting to dismantle the pillars that support democratic norms, and where resistance could be most helpful. It felt like the world was upside down, as many of us who usually criticize the media or the EPA for not being hard enough on industry suddenly found ourselves defending institutions that offered at least some path to accountability. Having listened to people at LMOGA strategize about removing impediments to their profit, I was less surprised than many of my progressive friends at the role big money played in the dismantling of our civic institutions.

My time in Louisiana helped me understand how ordinary, decent people could support not just the fossil fuel industry but those seeking to undermine government oversight. Everyone I met there was generous and hospitable, regardless of their politics. Louisianans are proud of their distinctive culture—from jazz and zydeco to gumbo and jambalaya—which developed from the mixing of African, European, Indigenous, and Caribbean cultures. Yet, it's also a state where the history of slavery is close to the surface, like the water table, so it is a particularly good place to explore the roots of our current racial divisions.

As I returned to the state four times over two years, I learned why ignoring race doesn't lead to equality, but I also learned why we shouldn't overemphasize what divides us. The Louisianans who were most successful in challenging powerful companies were often the ones who acknowledged inequality but emphasized that we share the air, water, and climate of this planet. Through their storytelling, they taught me much about finding common ground, and what gets in the way of it.

PART II

The Cost of Separation

CHAPTER THREE

WHAT'S RACE GOT TO DO WITH IT?

Margie Richard led a successful campaign against Shell, one of the most powerful companies in the world. By the time I heard her story in 2018, Margie was a grandmother in her seventies and the first African American winner of the prestigious Goldman Environmental Prize. Still, the events of previous decades were scarred into her memory. "Right here is where a young man was cutting grass," she said, pointing to a spot a stone's throw from the chemical plant in Norco, Louisiana. A spark from Leroy Jones' lawnmower ignited ethylene, which had leaked from a plant pipe. Neighbors heard the explosion and rushed outside to see Leroy running down the street engulfed in flames.

The sixteen-year-old died in the hospital. The woman whose grass he was cutting was killed instantly. Margie never forgot the sight of a sheet over Helen Washington's body or the stench of her singed hair. Afterward, she cried whenever she heard a lawnmower.

Margie was thirty-one and living nearby with her children at the time of the 1973 accident. As we drove my rented Nissan up and down the four straight, narrow streets of her old neighborhood, Margie pointed out a bald cypress and her favorite live oak, where children used to steal honey from a beehive. In the next breath, she pointed to the exposed pipes by the chemical plant fence, only feet from where her trailer used to stand. She explained who lived where in a community where people watched each other's children and checked on each other's elders. It was a mark of how bad the health effects of the chemical plant were that anyone wanted to leave.[71] Chemical emissions were so routine that children got rashes from wearing clothes that had dried on the line. If people noticed a chemical flare from the plant, they pulled their meat off the barbecue. Margie's sister, Naomi, started coughing up blood after she moved out. Initially misdiagnosed, she died in 1983 of sarcoidosis, an inflammation disorder associated with toxic exposure. Margie was devastated.

STORIES LIKE this are endemic in Cancer Alley. When I first heard of the region, I confess I imagined something closer to an actual alley—narrow, deserted, maybe with a dumpster and a dead end. That's a Philly girl's picture of an alley. In fact, the chemical corridor is massive, encompassing well over a hundred petrochemical plants on both sides of the Mississippi. It takes hours to drive if you follow the river with its wide oxbow turns. Between the belching plant stacks are working sugarcane fields, antebellum plantations preserved for tourists, and towns like Norco. Standing by the fence line with Margie, I grasped the power imbalance between a behemoth, global corporation and a small community of people, many whose ancestors were enslaved on this land, or close nearby. I could also feel the illusion of separation at work. A strip of tall trees and a railroad track were all that separated Margie's neighborhood, called Diamond, from the more populated white part of Norco, where the Shell oil refinery stood only a mile from the chemical plant.

In 1988, a corroded vapor line at the refinery caused an explosion that released 159 million pounds of chemical waste over both sides of town. Ceilings cracked in Diamond, but this time, those hit hardest were on the white side of Norco. Seven workers were killed, and forty-eight more people were injured, many when their homes blew apart. The refinery accident was much larger than the explosion in '73, large enough to be heard in New Orleans. But to Margie and her neighbors in Diamond, that didn't excuse the stark differences in how the two communities were treated.[72]

Shell paid over $40 million to the families of the seven men who were killed at the refinery. In contrast, a Shell employee slipped $500 to the mother of Leroy Jones for the loss of her son. The company paid nothing for the life of Helen Washington and only $3,000 to buy her property. Thirty years later, Shell was still commemorating the '88 accident, but company officials admitted that they didn't even have a record of the two deaths that traumatized people in Diamond. The media didn't cover those deaths either, while the accident on the white side of town garnered television coverage and stories in *The New York Times* and *The Washington Post*.

Feeling that their lives didn't matter to anyone else, people from Diamond formed a group to stand up to Shell. Known for her faith and forthrightness, Margie was elected leader. During the early years, they felt powerless and alone. Picketing on the levee outside the chemical plant with signs only brought dismissive comments from the workers. Although they were called the "Concerned Citizens of Norco," only people from Diamond seemed concerned about the chemicals they were all breathing. A turning point came when Margie saw a notice about a public session on the petrochemical industry and recruited a friend to go with her. There they met Dr. Beverly Wright, founder of the Deep South Center for Environmental Justice (DSCEJ), which harnesses academic research to support frontline communities. Dr Wright told Margie about people from other parts of Louisiana who were engaged in similar struggles.[73]

Margie started attending DSCEJ trainings and learned that the Concerned Citizens of Norco were part of a growing movement to address the disproportionate effects of environmental problems on people of color, a pattern called "environmental racism." "The pattern is all over the United States," she said.

THE ROLE OF RACIST POLICIES

I was told by several white Louisianans that there was no such thing as environmental racism because the people who worked for industry were "good people." Black people's proximity to industry was caused by poverty, not racism, they asserted. In fact, both race and class influence who lives closest to toxic facilities, but when compared directly, studies show that race correlates more strongly. Even the EPA during Donald Trump's first administration acknowledged that, though poor people in general breathe in more dangerous particulate matter, "results at national, state, and county scales all indicate that non-Whites tend to be burdened disproportionately to Whites."[74]

To better understand the role of race, I compared what I was learning about Cancer Alley to historian Ibram X. Kendi's definition of *racism*: "A marriage of racist policies and racist ideas that produces and normalizes racial inequalities."[75] Like the old saying, "Racism is prejudice plus power," this more precise definition dispels the common misconception that racism is just a matter of "bad people" who are prejudiced. Gradually I realized that the policies and ideas that created Cancer Alley all trace back in time. Margie clearly understood this, which is why she wove history throughout my tour of Norco.

FROM AN elevated spot by the river, with large ships behind us, Margie pointed out which plantations had been where. Each had river access used for the importation of enslaved Africans and the export of sugar. She told me she didn't like to dwell on the past. It makes her upset, wondering how people could have treated other people like that. But she is proud to be related to a participant in one of the largest rebellions of enslaved people in US history. It started just upriver in 1811 when hundreds of enslaved people marched toward New Orleans, passing the street where Margie used to live. After plantation owners crushed the rebellion, with help from a US militia, they displayed sixty-three decapitated heads on stakes along River Road as a warning. "If my ancestors can have their heads put on stakes and fight against slavery, I can surely stand up to Shell," Margie said.[76]

After the Civil War, the newly emancipated asserted their freedom. Fearful of losing their workforce, plantation owners adopted new policies that made it hard for them to buy land, leaving many to work the same fields on which they had been enslaved, now as sharecroppers. In the early twentieth century, when oil companies bought up plantations, seeking the same river access, many of the lanes between them were still inhabited by Black families who had been there for generations. An oil refinery was built on the old Good Hope Plantation, just downriver from the Diamond Plantation. The post office renamed the whole area Norco, an acronym for the New Orleans Refining Company, and a sign of the region's shifting identity. Shell purchased the refinery in 1929 and attracted white workers to Norco with company housing.

By the 1950s, Shell wanted a chemical plant to process refinery by-products for newly popular synthetic materials. "Believe it or not, I grew up right over there," Margie said, pointing toward land paved over by the chemical plant. In a small community called Belltown, her grandfather had been an independent farmer, producing pork, beef, cantaloupe, and corn. Her father kept plum, peach, and fig trees, as well as chickens and rabbits. Despite being warned about water moccasins and diamondback rattlers, Margie and her sister Naomi would sneak down to the Bonnet Carré Spillway, built to divert water from the Mississippi River to prevent flooding in New Orleans. They would pick blackberries and mix them with sugar and ice. "Ooh, it was so good!" Margie recalled with a grin. I tried to imagine the vibrant land she described before the residents of Belltown were told they had to move with no explanation and meager, if any, compensation.

Prior to the campaign wins of the freedom movement in the mid-1960s, many Black Southerners were threatened with violence if they tried to vote, let alone resist a chemical plant. Strict racial segregation limited their housing options, so people from Belltown bought or rented plots on the land right next to the new plant, still called Diamond after the old plantation. To me, the distance from Belltown to Diamond seemed short, maybe a few city blocks, but the move meant that Margie's family lost the soil and trees they had cultivated for generations. In Belltown, growing food gave them some economic independence. In Diamond, they were dependent on wage labor, even though few of the new plant jobs went to Black workers.

For those with land to sell, Louisiana law is complicated. It is based on the Napoleonic Code, which requires sellers to show clear title back through the generations. Most descendants of enslaved people can't produce that kind of paperwork for property that was passed down without going through probate. The Napoleonic Code also requires that in this situation the proceeds of a home sale must be shared among all the family members descended from the original owner, which Margie explained might include many cousins. Even though the law doesn't mention race, it disproportionately burdens Black Louisianans who want to move away from industry.

WHILE SOME aspects of Margie's story are particular to Louisiana, racial segregation has a long history across the United States, shaped by an array of racist policies. When US agencies first appraised neighborhoods after the Great Depression, they undervalued property in communities of color. Banks used the red lines government drew to exclude neighborhoods where they would not lend, or would only lend at high interest rates. Recent research finds a strong correlation between communities that were redlined in the 1930s and those that became sites of fossil fuels facilities in the following decades.[77]

Cheap land is part of what industry looks for when building a new facility, so whether or not executives are personally prejudiced, their companies benefit from policies that have devalued Black- or Brown-owned property. These same policies have made it harder for people of color to build enough home equity to sell their homes once there is a toxic facility next door. This national pattern also affected Diamond. Some Diamond families sold their homes to Shell over the years for an average price of $26,933, while homes on the white side of Norco had market values four times that, even though those homes were not that much bigger and were also close to industry.[78]

Given all these obstacles, the Concerned Citizens of Norco decided to focus on getting Shell to buy them out at prices high enough to enable them to relocate. Because of racist policies, this was intrinsically more difficult for them than for their white neighbors. Because of racist ideas, those neighbors alleged that

Diamond residents were "just out for a buck," or wanted "something for nothing."

THE ROLE OF RACIST IDEAS

Even when racist policies are dismantled, racist ideas can maintain and justify the inequality those policies created. For example, the stereotype that many Black people are criminals can influence the behavior of real estate appraisers, realtors, and home buyers, contributing directly to disparities in property values, even for neighborhoods with low-crime rates. The resulting low tax base creates underfunded schools, which lowers property values still further. This cycle makes it easier for industry to move into a neighborhood and buy off politicians with a modest donation to the local school board.

Ibram X. Kendi shows how stereotypes were deliberately created to justify the lucrative system of slavery and were then adapted through the centuries to support changing interests. In *Stamped from the Beginning: The Definitive History of Racist Ideas in America,* he writes, "In all of the fifty suspected or actual slave revolts reported in newspapers during the American colonial era, resisting Africans were nearly always cast as violent criminals, not people reacting to enslavers' regular brutality, or pressing for the most basic human desire: freedom."[79] By promoting the criminal stereotype, the media served as a pillar supporting slavery. Religion served as another pillar when preachers claimed that slavery was God's will.

I was struck how many old stereotypes were deployed to discredit the Concerned Citizens of Norco. A Shell spokesperson told an interviewer that, as the campaign grew, the company was "worried about some outsider coming in and stirring the crowd and creating something almost like a riot atmosphere."[80] Never mind that Margie was a churchgoing teacher and mother motivated by a simple desire to protect the people she loved. She told me with an eye roll that their meetings often drew a heavy police presence.

I CAME to Louisiana knowing that there were racial disparities in people's exposure to pollution. I soon realized that environmental injustice was also about how people are seen and treated—by the press, industry, and different levels of government. Whose symptoms are believed? Whose lives are valued? Margie was frustrated that people in Diamond had no guaranteed evacuation route. Between the chemical plant and the refinery, accidents happen. That risk is compounded by the fact that Diamond is boxed in by train tracks, the river, and the spillway, which diverts water into the floodplain when the Mississippi is high. "You get it?" she asked when I was slow to grasp the problem. "In the worst case scenario—which happened!—when the water is here, you can't go that way. If a train is on the track there, you can't go there." During one accident, people in Diamond had to evacuate through the narrow streets of the white neighborhood, making them last to escape any dangerous exposure. They raised this issue with Shell, but the company said it was not their problem.

It is hard to prove the role racist ideas play in corporate or government indifference. In *Toxic Communities: Environmental Racism, Industrial Pollution, and Residential Mobility*, environmental sociologist Dorceta Taylor notes that communities must prove "*intent* to discriminate" if they want any help from the courts. That is extremely difficult, even when there is a pattern of disparate treatment. Taylor cites one study that showed that oil refineries that violated environmental laws were fined the most in white census tracts, significantly less in Hispanic ones, still less in Black tracts, with the lowest fines reserved for communities with the highest percentage of Black people.[81] This devaluing of Black lives is not a matter of official policy, but Margie says the patterns are clear.

IN RECENT decades, psychologists have come to understand that many of our racist ideas and behaviors are unconscious. I first became aware of my own unconscious racism in 1987, almost two decades before learning the term "implicit bias." I was walking at dusk in downtown Philadelphia and noticed a young Black man walking slightly behind me. Reflexively my hand tightened around my purse strap, which I observed with surprise. I had just returned from two and a half years living in an African village, where I never felt unsafe. Yet here I was replicating a racist reflex I had criticized in my mother. Not wanting to think of myself as racist but curious if this was a pattern, I started paying attention to how my hand reacted to people of different races. I noticed that it only twitched slightly near Black men in suits or professionally dressed Black women. An African man in a dashiki did not provoke a purse grab at all. My unconscious reaction showed up around those whose image was most demonized on television: young, urban Black men. I also noticed that once I started paying attention, my hand's reflex became much weaker.

In the decades since, I've realized how toxic ingrained fear can be. We are less likely to feel compassion for people we were taught to fear. We more easily suspect them of unsavory or even hostile motives. Implicit bias tests have shown that most American minds automatically link African Americans and weapons, even when the weapons depicted in the test are bows and arrows. While whites are statistically most likely to make this association, it is also shared by many Black, Latino, and Asian people. This is why hiring diverse law enforcement, while important, is not enough to end unequal police violence.

I suspect that the unconscious nature of much prejudice contributes to the perception gap that often occurs between racial groups. People of color see the unfair way they are treated, while white people deny that discrimination was intended. During a 1998 Shell chemical plant accident, the school bus for kindergarteners was diverted to a different school, alarming parents who weren't informed. Amid pungent smells, some rushed toward the accident searching for their children. In a follow-up meeting at the American Legion, residents from both sides of Norco were asked to raise their hands if they had received a phone call from the company warning them about the accident. Only white people raised their hands. When Black people pointed out that they hadn't been called, some whites accused them of "trying to make this a race issue,"[82] a common tactic for discounting allegations of racism.

PROVING THE PATTERNS

"This is not for wimps," warned Dr. Robert Bullard, recounting how much pushback academics like himself faced when their research first showed the correlation between race and pollution. Often referred to as the "father of environmental justice," Bullard was speaking to three hundred students and faculty from Historically Black Colleges and Universities (HBCUs), who had gathered in New Orleans for the sixth annual HBCU Climate Change Conference. "Some of these government agencies, they want to think they are dealing with people who are afraid," he told the students. "But many of us came out of the civil rights movement, where we were not afraid of the police or dogs, and certainly aren't afraid of some government bureaucrat."

The HBCUs include Howard, Morehouse, and Spelman, as well as schools I was less familiar with, like Southern, the General's alma mater, and Xavier, where we were meeting. HBCUs played a vital role during the freedom movement. The lunch counter sit-ins, the Freedom Rides, the march from Selma to Montgomery—many of the iconic events of that era were led by students and alumni. Likewise, HBCU alumni and faculty played prominent roles in establishing the environmental justice movement in the 1980s and '90s. Many of those founders were at the 2018 conference, which was co-chaired by Dr. Bullard and Dr. Beverly Wright, who played a key role in helping Margie and the Concerned Citizens of Norco plug into this wider movement.

Vernice Miller-Travis, who was also part of the Norco campaign, took the stage to share her experience publishing the groundbreaking study *Toxic Wastes and Race in the United States* in 1987. With the support of the United Church of Christ, she took the EPA's list of Superfund sites and painstakingly mapped them across every county of the United States, then compared them with census data. The map showed red dots across the country with dramatic concentrations in Black and Indigenous communities in the South and Southwest.[83] Detractors insisted the data must be wrong. Industry commissioned vicious articles personally attacking Miller-Travis. "It's amazing the amount of effort put into discounting the role that race plays," said fellow panelist Paul Mohai, one of the few white academics present at the conference.

MOHAI FELT the role of race was well documented in the research, but that more work was needed to prove the link between pollution exposure and health disparities. This issue is particularly complex in a region like Cancer Alley, where there are so many industries emitting so many different chemicals into the wind. Margie's sister died at age forty-three of sarcoidosis, which averages one in a thousand people, yet Margie knew of three other people with the disease in her tiny community.[84] She'd visited other frontline communities with high rates of sarcoidosis as well as other diseases, like kidney failure and cancer. To her, the pattern was clear. Living next to industry makes people sick. Yet, showing a pattern is not the same as scientifically proving that a specific person's illness was caused by emissions from a specific company.

I heard many white Louisianans blame the state's high cancer rates on individual choices, like smoking and eating fatty foods, rather than the systemic problem of industry pollution. While it's true that cigarettes and fried foods increase cancer risk along with other factors, like genetics and age, that explanation often felt racially charged when I heard it used to dismiss the concerns of Black people. One scientist I met, who was doing contract work for Shell, implied that the people who live near polluting industries were too stupid to understand science.

I asked biologist Kimberly Terrell about the complexity of cancer risk and appreciated her framing. A white woman who began her career studying wildlife, Terrell's job at the Tulane Environmental Law Clinic now involves supporting communities in Cancer Alley. She noted that industry likes to portray things as either/or, when in fact many things can raise one's risk, including chemical exposure and the length of that exposure. "Things that you breathe in that are known to cause lung cancer are going to increase your risk of getting lung cancer. That's undeniable. Whether or not that translates into an outcome of lung cancer is a secondary question," she explained. She said it is unfair for some communities to live with greater cancer risk, and that should be the focus of the argument, not how much cancer is caused by chemicals and how much by diet or other factors. Industry should have to prove that their emissions are safe rather than expecting communities to prove they are dangerous.

That is not how the burden of proof works, however, especially in court. When 250 Diamond residents brought a suit against Shell, they argued that the facility "hurt their health, diminished the value of their property, and made them fear for their safety," an article in *Legal Affairs* summarized. Many told personal stories of illness and loss. Company experts contested their claims and brought in whites from Norco, many with ties to Shell, to testify that they were not bothered by pollution. During closing arguments, opposing council implied that all Diamond residents cared about was profiting from the situation. Paraphrasing the 1996 film *Jerry Maguire*, Shell's lawyer said, "Before you show them the money, they must show you the proof." The jury sided with Shell and against the people of Diamond. Losing the lawsuit was a devastating setback for the Concerned Citizens of Norco, half of whom sat in the courtroom every day of the proceedings. At that point, Margie was tempted to give up, but in the middle of the night she felt what she described as the Holy Spirit encouraging her to continue.[85]

ALONG WITH an insistence on academic rigor, the HBCU climate conference included explicit references to faith and spirituality. Several people mentioned the importance of getting rid of ego and valuing something more than making money. At the same time, this was a group that was not naive about power. During one Q and A, an older man in the audience stood at the microphone and said, "What's that they say? If you're not . . ." The room finished his sentence: "at the table, you're on the menu." A woman in the audience added that they needed to be part of setting the table and planning the menu, which met with claps and mm-hmms.

Many conference presenters used the pronouns we and ours when referring to the communities most impacted by environmental racism and climate change, which were understood to be deeply interrelated. Talking about the aftermath of Hurricane Katrina, Dr. Wright described the damage to her own home, which I had heard became a refuge for her staff, whose homes were completely gone. A panelist who was from Chester, Pennsylvania, talked about crying as she watched the Katrina disaster unfold on television, an experience that helped lead to her career working on international climate policy. One student presented data showing that his coastal college would become underwater due to sea-level rise. Other speakers were from Houston, which was recovering from Hurricane Harvey. In contrast to LMOGA, no one at the HBCU conference was under the illusion that they were separate from the subjects they were discussing. In fact, the year's theme was "Fighting for Our Lives."

At its heart, the work of these academics/advocates seemed to be what Dr. King called "love implementing the demands of justice."

THE HANGOVER OF HISTORY

To help the students understand the lineage of resistance they were part of, the HBCU Climate Change Conference took a field trip to the Whitney Plantation, on the other side of the river from Norco. On the bus ride, Margie talked about Diamond's struggle, and how they finally got out of Shell's shadow. When we stepped onto the soil of the old sugarcane plantation, the students fell quiet, as if we were in a church or a cemetery.

Unlike nearby plantation tours, which celebrate the glamorous lives of the enslavers, the Whitney focuses on the lives of the enslaved, what they suffered, and how they resisted. Our tour guide, Yvonne Therese Holden stressed that Africans were enslaved in part because of the skills they brought in agriculture, building, and trades like blacksmithing. They were the ones who built the beautiful large homes featured on plantation tours. "When a person worked slowly or pretended not to be able to figure out a certain chore, that was a way of resisting," she explained as the students nodded solemnly. "When you hear a stereotype that Black people are lazy or stupid, you have to realize that people were doing this on purpose."

As we walked the eerie, beautiful grounds, we paused in front of large metal bowls used to process sugar, which had to be done quickly to avoid sugarcane rot. The Whitney's director of operations, Yvonne explained that enslavers tried to divide the people they kept captive by making Black men drivers, those who enforced the brutal work pace. When we moved into the two-story "Big House," she acknowledged that women were raped there, dispelling the myth that those who worked inside had it cushy compared to those in the fields. Tall with tan skin and pulled-back curly hair, Yvonne said that colorism—the idea that light-skinned people of African descent are superior to their darker-skinned kin—was yet another way of dividing people.

Later when I ran into Yvonne in New Orleans, she told me that when Thomas Jefferson purchased the Louisiana territory from France in 1803, Americans were shocked to find so many affluent, skilled, and educated free people of color. Mixed-raced people were an inherent threat because they punctured the illusion of clear indelible lines between the races. The United States soon imposed laws of strict racial separation like those already common in the rest of the South.

THE HISTORY of Whitney Plantation was uncovered by Formosa. In the early 1990s, the Taiwanese plastics company wanted to build a rayon plant on the site, but it met fierce resistance from a combination of preservationists, environmentalists, and community members. Hoping to divide the opposition by placating the preservationists, Formosa commissioned a study of the plantation's history and promised to build a small museum. Opposition to the new plant continued, including civil disobedience and a march that got media attention. With rayon fading in popularity, Formosa gave up and sold the property to John Cummings, a wealthy white Louisiana lawyer. Cummings read the eight-volume history Formosa commissioned and was so moved, he decided to turn the property into a slavery museum. "I started to see slavery and the hangover from slavery everywhere I looked," Cummings told *The New York Times*.[86]

The hangover is national, but it's unabashed in Louisiana. Upriver from Norco, St. James Parish published an economic development plan in 2014 that was overtly nostalgic for the days of plantation slavery. To augment its case for new industrial development, the document quotes a book published in 1957: "The early 1800s was the era of fabulous plantation life in St. James. Acreage was counted by thousands and slaves by hundreds. It was the day of luxurious living, of sumptuous entertainment, of delightful ease. Sugar was gold."[87] The implication was that petrochemicals could replace the wealth built from sugar—with no recognition that "plantation life" brutalized and killed people and only the plantation owners got wealthy from it. In the five years following this plan's publication, St. James became the terminus of the controversial Bayou Bridge Pipeline and the proposed site of Formosa's new $9.4 billion plastics plant, which the neighboring Black community was fighting to stop.

Back at the HBCU conference, several speakers connected the dots between the field trip to the Whitney and the corporations that had taken over many plantations. A few spoke of the Indigenous people who stewarded the land before it was stolen, a foundational aspect of the history I would learn more about in other states. This was another aspect of the illusion of separation, I realized, the illusion that we are separate from our history. The evidence was in the soil itself, depleted of nutrients from the growing of sugarcane long before it was polluted by the processing of oil and its by-products.

That afternoon, the conference honored Margie Richard for her leadership resisting Shell. Several people in the large room had supported her as that campaign grew. "Thank God for the Deep South Center for Environmental Justice," Margie told me.

FROM NORCO TO NIGERIA

The Concerned Citizens of Norco attracted more allies. Attorney Monique Harden worked for Earthjustice Legal Defense Fund before joining DSCEJ staff. Anne Rolfes founded the Louisiana Bucket Brigade, named for a system of using buckets to collect polluted air, something residents could do themselves to establish which chemicals were being emitted. MacArthur Award-winning chemist Wilma Subra had long shared her expertise with Diamond and other communities. She did the testing and confirmed that many of the chemicals being released were known or suspected carcinogens.[88] These efforts were especially important after the '98 accident, when air samples collected in Diamond directly refuted the company's claims. By then, larger groups were amplifying the community's demand for relocation, including Sierra Club and Greenpeace.

Meanwhile, Shell was facing global criticism for its human rights abuses in Nigeria. The company had extracted oil in the densely populated Niger Delta since the late 1950s, severely polluting the water and soil that subsistence farmers and fishermen relied on to feed their families. The most affected were the minority Ogoni people, who in the early 1990s mobilized a nonviolent movement to protect their "Shell-shocked" land. The Nigerian government—which received a large percentage of its revenue from oil—responded by killing and maiming hundreds of activists and hanging nine leaders. Evidence indicated that Shell actively encouraged, and in some cases paid for, this brutal repression. One month after the executions, Shell signed an agreement with Nigeria to invest $4 billion in a liquefied natural gas plant. Fourteen years later, in the face of international grassroots pressure, Shell agreed to a financial settlement with the families of those who had been hung.[89]

Shell tried to rehabilitate its image, rebranding itself and donating millions to environmental causes. Leaders of the Norco campaign realized that this gave them leverage, essential for a tiny community taking on a giant company. They began explicitly linking Shell's greed and destruction from the Mississippi Delta to the Niger Delta, even using the phrase "Norco to Nigeria" and threatening to embarrass Shell at international conferences.

AS WE drove slowly through Diamond, where grass has replaced most houses, Margie reminisced about her experience representing

Diamond at the United Nations Commission on Human Rights in Geneva in 1999. Dr. Wright and Monique Harden coached Margie on her two-minute time limit, but Margie insisted on taking a break for a cup of tea. Wondering what in the world she was doing in Switzerland, she remembered a hymn that said the Lord could speak through her. After that, she felt peaceful. When the time for her testimony came, she realized she was there to represent common people, unlike the professional representatives of so many organizations. "I put my paper down, and I spoke from my heart," she said. She felt God gave her the words. Afterward, the head of Amnesty International asked for a copy of her speech. Laughing, Margie recalled saying, "Sir, if you didn't record it, I don't have it."

That led to the first breakthrough. In 2000, Shell offered to buy out the two Diamond streets closest to the plant. People in the close-knit community wanted the offer to extend to all four streets, and accused the company of intentionally causing division and resentment among them. Organizers worked hard to keep the residents united, drafting demands that would meet the needs of those who wanted to stay, as well as those who wanted to move—owners and renters alike. To keep the pressure on Shell, they also hosted many creative actions, such as challenging the company to a debate to coincide with the presidential debate. Shell didn't show, a fact that made the news, as did a pray-in at Shell's US headquarters in Houston.

Margie felt God at work again in 2000 when international climate treaty negotiations were held in the Netherlands, Shell's global headquarters. Margie was not scheduled to speak at a gathering that included company executives, but she raised her hand and prayed they would call on her, even though she was shorter than the people around her. When they did, she brought forward a bucket of air from Diamond and a vile of polluted water from Nigeria and invited the executives to breathe and drink the samples. It was a powerful symbolic way of exposing their illusion of separation. Then, she issued a passionate cry for help on behalf of her community.[90]

In 2002, Shell relented and offered to buy out anyone in Diamond who wanted to sell, or pay moving and utility down payments for renters who wanted to move. For those who chose to stay, the company offered home renovation funds. The company also promised to reduce emissions by 30 percent and invest $5 million in a health clinic that Margie proudly showed me, along with ambient air monitors, so the few who chose to remain could see when air pollution was high. The white residents of Norco also benefited from those monitors and the pollution reduction that the campaign achieved.

Near the end of my afternoon with Margie, we crossed the line of trees separating Diamond from the white side of Norco. Slightly bigger with more brick than wood, the houses were still populated, and there were Virgin Mary statues on several lawns. Margie said a few African Americans lived here now, but not many. I speculated that the people on these streets must have health problems from pollution, too. "They do!" she said. "You can't tell the wind where to blow. The same air I breathe, you breathe."

I smiled and told her about the words that had come to me a few years earlier: "The crisis of the Earth is saving us from our illusion of separation."

"I agree with you. That's why I let you take me around in your car," she said with a laugh. "I see one race, the human race. We all need air, water, and land. We are messing ourselves up by not protecting what God created," she concluded.

LESSONS FROM NORCO

South African Black Consciousness leader Steve Biko once said that the concept of race was invented by white people to justify their own greed, but over time it took on a life of its own. That's as good a summary as I've heard. I often imagine this life-form as a hydra, the multiheaded serpent of Greek mythology. That monster was especially hard to fight, since each head breathed fire. Likewise, racism has different manifestations, which shift and adapt over time. Each is dangerous on its own but lethal when discriminatory policies and prejudiced ideas protect each other. The hydra of racism defends the very companies contributing the most to climate change.

Racism has kept property values low and helped industry defeat legitimate lawsuits. Whether it's spewing chemicals in Cancer Alley, dumping fracking wastewater in Mexican American communities in Texas, routing pipelines through Indigenous land in the Dakotas, or building gas plants like the one near the Vietnamese community of New Orleans East, the fossil fuel industry has gotten away with endangering people's lives at least in part because the people at the most immediate risk are the least valued. Racism has essentially operated as a pillar supporting industry's power.

Many heads of the hydra sprouted from slavery—not only in Louisiana, where the connection is embedded in the land, but also nationally through the racist ideas invented to justify the capture and exploitation of Black bodies. Stereotypes suggesting that Black people are lazy, criminal, and responsible for their own bad fortune are still used to blame them for racial inequality. Making these connections helped me understand why the illusionists of today are so determined to ban this history from school curricula and libraries, especially the books of Ibram X. Kendi, who is one of the most banned and maligned authors in the United States. If we all understood the importance of policies and ideas, we might realize that policies and ideas are things we can collectively change.

THERE IS deep resistance to this learning, and not just from those politicians trying to defund our education system. Many years ago, I began asking questions about the history of slavery at historic sites near my home in Philadelphia. A shocking number of tour guides told me that there was no slavery in Pennsylvania because of the Quakers. I responded that I was a Quaker, and while Quaker abolitionists contributed to Pennsylvania passing a gradual abolition act in 1780, Quakers had enslaved people in the colonial period. One older female docent responded that maybe slavery "wasn't so bad if the owners were nice." I said I didn't think there was a nice way to enslave another human being, and she averted her eyes.

When touring Whitney Plantation, I noticed that, like the Pennsylvania docent, most of the visitors were white. I asked Yvonne what she thought helped white people to take in this difficult history, and what might block them. She took a deep breath and said, "Good question." She noted that many white people identify with the enslavers on some level, so it's hard for them to hear how brutal and greedy they were. "There must have been some good people," was a comment she heard often. "They want to think of themselves as good, and it's difficult to reconcile that with the bad things these other white people did." I told her that I'd heard several Louisianans say that there couldn't be such a thing as environmental racism because the people who worked for industry were "good people." She agreed there was a parallel. She found it helpful to say at the end of her tours, "We didn't choose this history. It's a shared inheritance. So, what do we do with that inheritance?"

It may be psychologically easier to discount those revealing a system's cruelty if we have benefited from that system. Or we may avoid thinking about systems altogether, which is easy if we grew up in an individualistic culture that overwhelmingly tells us that individuals shape their own lives. While this message is pervasive in the United States, people of color are more likely to understand how systems affect them, given such a clear view of the system's dirty underbelly. The illusion of separation operates like a veil for whites, keeping us from realizing that our own children's asthma is caused by a pollution-spewing facility a neighborhood or two away.

Donald Trump's 2016 election put a spotlight on white working-class people, like the Louisianans who sided with industry even though they, too, were being sickened by pollution and battered by hurricanes. On the final day of the 2018 HBCU climate conference, Dr. Beverly Wright sincerely asked the audience, "How can we reach them?" I shared Dr. Wright's question. New fossil fuel infrastructure is only politically feasible when whole communities are deemed expendable and when those who might find common ground with them are divided. This continued to be true under the administration of Joe Biden, who championed renewable energy, but didn't stop fossil fuel expansion. The role of racism as a dividing force felt even clearer after the 2024 election.

One of the Norco images that haunted me was the old strip of trees that divided Diamond from the white side of town, where people were still deeply invested in the illusion. Margie knew that they all breathed the same air, but somehow her white neighbors didn't. While the Concerned Citizens of Norco won their campaign without those neighbors, they only escaped the flaring chemical plant by a few miles. They didn't shut it down. The chemical plant emissions are still contributing to regional pollution, as well as to climate change. I couldn't help remembering that, according to legend, killing the hydra was the only one of Hercules's twelve challenges that the hero could not accomplish alone. When we all acknowledge our shared stake in clear air, clean water, and a stable climate, and join our power with the neighbors just over the tree line, it can shift the balance of power between companies and communities.

Of course, the hydra will defend itself, so we have to have each other's backs.

CHAPTER FOUR

DIVIDING THE SPECTRUM OF ALLIES

The General suggested that we visit Mike Tritico, who had fought toxic pollution in southwestern Louisiana since the 1970s. As soon as Mike heard me mention the illusion of separation and the question of what helps people to find common ground, he pulled out a crumpled piece of paper and started drawing a map. Ground zero was Willow Springs, a small African American farming community where the petrochemical industry started dumping its toxic waste in the late 1960s.

"Right across the river from Willow Springs was a rich, white neighborhood. This way was not as rich, but white," said Mike, the descendant of Italian immigrants. "That way was more like hillbillies. So, there was all this variety of people. Everybody was suffering together." Over time, people from those different communities teamed up and built enough political power to dramatically reduce the harm the dump was causing. Mike said it was a great model of everyone coming together. Later, a spy used racism to drive them apart.

I was intrigued. I'd heard that Calcasieu Parish had an even higher concentration of petrochemical facilities than Cancer Alley. Companies were attracted by raw materials like petroleum and salt, as well as a ship channel to the Gulf of Mexico and a highway that ran from Florida to California. We were meeting not far from Mossville, a formerly vibrant Black community that is now the site of a giant petrochemical complex. In both Willow Springs and Mossville, Black people suffered the most, but in the first case, many whites joined a campaign that improved conditions for everyone. In the more recent story, most whites sided with industry, and industry won. The resulting Sasol complex was not only a "super polluter," it released greenhouse gases in huge amounts.[91]

I was curious what these two stories could teach those of us looking for common ground today. To visit the place where the dump struggle started, Mike and I climbed into the General's Land Cruiser, along with Mike's friend, Peggy Frankland, who joined the campaign after realizing the dump was four miles from her own middle-class home.

SITTING ON his patio under a row of magnolias, Herbert Rigmaiden leaned forward in his overalls, white hair sticking out from under his baseball cap. Pines had fallen in a recent storm on this farm, which Herbert had worked since his childhood in the 1930s and '40s. Named Willow Springs for the clear, fresh water that used to bubble to the surface, the nearby woods provided deer and geese as well as wild hazelnuts, hickory nuts, blackberries, and mayhaws, a local fruit popular in jelly. These gifts of the land supplemented the crops and animals that Herbert's African American family tended. A river and woods shielded them from the wider world, or seemed to. No one informed them when the land next door became the petrochemical dumping ground for Calcasieu Parish, supposedly named after an Atakapa chief whose name meant "Crying Eagle."[92]

At the farm's height, Herbert's family owned a hundred head of cattle. "I think we had about thirty-two cows come up with sores on them about that big," he recalled, holding up his strong hands to approximate the gaping wounds, oozing from chemical exposure. When the cows couldn't be saved, Herbert cut some open. He found green goo that smelled like the chemicals at the dump.

"And the cows that didn't die; he was sending them to the stock!" interjected Peggy.

"You probably ate some of the meat, all of y'all," said Herbert.

I asked where his meat would get shipped, and he said, "All over the world, baby."

That's what Herbert told a congressional committee when he and Peggy traveled to Washington, DC, in 1984. "Every time you put a beef steak up to your mouth, just remember, it could have come off of my farm!" Peggy recounted.

WHILE THOSE closest to toxic facilities are disproportionately low-income and/or communities of color, pollution spreads through air, water, and food, ultimately affecting everyone. Some believe that industries prefer to locate toxic facilities in places where the neighbors are divided by race, class, ethnicity, or even by county or state lines, on the theory that people are unlikely to mount a joint opposition across such lines. This theory rang true to Minh Nguyen, who felt that the mixture of white, Black, Asian, and Latino people in New Orleans East was part of what made them a target. Racial division certainly worked in Shell's favor in Norco, a few hours' drive to the east of Willow Springs. Calcasieu Parish was also starkly segregated.

Herbert's community suffered from noxious fumes, dying animals, and a dramatic increase in illness for a decade before people on the opposite bank of the narrow river even noticed. That didn't mean whites were unaffected. The dump stored, and sometimes burned, a host of toxins, including trichloroethylene (TCE), polychlorinated biphenyls (PCBs), and vinyl chloroform, not to mention lead and mercury. Open, unlined pits drained right into the Chicot Aquifer, the drinking water source for millions of people. When it rained, the pits often overflowed into the Little River, which flowed into the West Fork of the Calcasieu River, which passed through larger and more affluent Lake Charles. Eventually the dump grew to eighty acres and accepted hazardous industrial waste from all over the United States. By the time people in Lake Charles became concerned, many were sick or dying, though with so much industry surrounding them, it was hard to know from what.

When the different communities finally started to work together, they were able to drastically curtail the dump's operation, ending its use of dangerous open pits and getting some of the soil remediated. In addition to improving the air quality in Willow Springs, their campaign led to a statewide law on toxic dumping.[93]

"What we got done, we got because we all worked together," recalled Mike as we sat under Herbert's magnolias.

"That's right," nodded Herbert, leaning back in his chair. "Absolutely. Because if you don't work together, you ain't gonna get nothin' done."

THE POWER OF WORKING TOGETHER

Mabel Rigmaiden Jones, Herbert's sister, was the first to try to change things. "A lot of people dismissed her," confided Peggy as we were on the way to Mabel's house. When we pulled up, Mabel walked out to greet us, tall, broad, and in her mid-eighties. The two women lingered, holding hands. Mabel's daughter, Caroline, had recently died with a thyroid so enlarged she couldn't swallow the radiation pills. Her first doctor told her it was just a sore throat, delaying an accurate diagnosis. It was the first time Peggy had seen Mabel since Caroline's passing.

Leaning back under a large awning, Mabel described the air from the dump's open pits. "It would cut your breath," she said. Sometimes the family had to stay inside and close the windows despite the sweltering Louisiana heat. Mabel worried about her children, who walked on the country roads where dump trucks often spilled hazardous liquid. In 1968, she took a petition against the dump door-to-door while her children were in school. Although many of her neighbors were afraid to stir up trouble, she delivered five hundred signatures to the parish government. When they were mysteriously "lost," she canvassed her community again.[94] Meanwhile, many people in Willow Springs had started to get sick. Chickens died, and vegetables in the garden withered. Even Herbert's new truck corroded from the toxic air.

Ruth Shepherd was the first white person to get involved, one of many friends who had since died. The daughter of a Missouri farmer who taught her the name of every tree, she moved to Louisiana with her husband, a locomotive fireman. When she happened upon Willow Springs one day in 1977, Ruth became suspicious of the many trucks on the narrow road. Talking to Herbert the next day confirmed her hunch that there was a toxic dump two and a half miles from her home. Inspired by a national surge in grassroots environmental activism, she complained about the dump to a few public officials. When that changed nothing, she reached out to Mike Tritico, who had a local radio show on environmental issues.

Together, the Rigmaidens—Mabel, Herbert, their brother, and their mother—teamed up with Ruth, Mike, and a few others to form one of the first multiracial environmental efforts in the state.[95]

One of their early meetings drew two hundred people to the First Baptist Church of Willow Springs. Soon after, the national company that had purchased the dump gave the church pastor a $200 donation in exchange for his agreement that the church would not host any more activist meetings.

IT WAS five more years before Peggy learned about the dump. She read in the paper that chemical waste from an accident on the eastern side of the state was being shipped to her region. While her children were in school, she went looking for Willow Springs and was shocked by the foul odor and how close some wooden-frame houses were to the dump. "The more I saw, the angrier I got," she recalled.[96] Peggy had noticed an unusual number of illnesses in her white community. Curious if there was a connection, she returned to Willow Springs with a friend to do a health survey. To their horror, they found that every family within a mile of the dump had severe, often life-threatening medical problems. Peggy threw herself into the campaign to close the dump and was mentored by the more experienced women.

Over the years, they tried a variety of tactics, from advocating for scientific studies to confronting decision-makers. One time three of the white women brought water from the Rigmaidens' well and offered it in a champagne bottle to a Supreme Court justice in New Orleans. At a hearing, another group member lifted her dress, warning the commissioners that she was going to show them something they'd never forget. The men covered their eyes, but she only revealed her leg braces. "I have been diagnosed with chemical neuropathy," she explained. "I can't feel my legs. This is what happens when you issue permits. Now are you men ever going to forget this?" "No, ma'am!" the commissioners said.

When a truck driver was killed in another part of the state

from breathing in fumes after dumping a toxic load, the activists borrowed a coffin and held a symbolic funeral procession in solidarity. When they were blocked by a row of armed police on the only road into Willow Springs, a white Baptist minister de-escalated the situation by standing on the hood of a truck and preaching.[97] It wasn't the only time they faced danger. One day Mike and Herbert went out in a boat after a storm to inspect the dump ponds. They caught company workers dumping directly into the river. Between the company men, the wind, and the toxic fumes, they worried for their lives. When I later asked Peggy if they had received death threats, she laughed bitterly. They all had.

As the old friends reminisced, Peggy mentioned the outdoor meals they used to share and Mabel's delicious fig cake. The General, who had been listening quietly, observed that Mabel had a fig tree in her yard, not far from where we sat. She said she brought it with her when she moved five miles south of Willow Springs after her mother died of breast cancer. She couldn't carry big pots like she used to, but Mabel still loved to make a good gumbo. I smiled and asked if food was the key to bringing people together. "Oh, yeah," she nodded. "All I had to say was, 'I'm cookin'!'"

THE SPECTRUM OF ALLIES

The spectrum of allies is a tool to help movements build power. When teaching this theory, trainers often draw the top half of a pie and divide it into five slices. Moving from left to right, they label the slices "active allies," "passive allies," "neutrals," "passive opposition," and "active opposition." It's tempting to focus only on the two ends of the spectrum, active allies and active opposition, but what people closer to the center do is also crucial to making change.

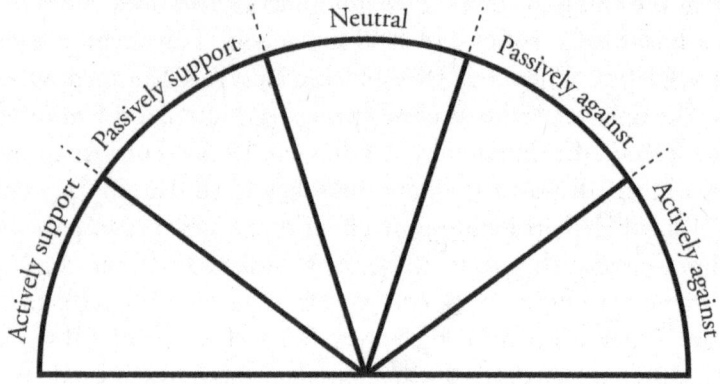

To use this tool, campaigners fill in the diagram with the names of specific groups or constituencies, then brainstorm strategies to influence those who might be moved closer to the activist position. That might mean inviting environmentalists who are already sympathetic to become active by showing up to a hearing or nonviolent direct action. Those who are neutral or not following the issue may need to be educated about the dangers of toxins, so they become passive allies. Getting media to show that industry is lying might prompt some who are pro-industry to at least loosen their allegiance. If each group shifts a little, and supporters become active rather than passive, that can build pressure on power holders and weaken support for the status quo. All those things happened in the Willow Springs campaign.

One ally who joined in the final years was Shirley Goldsmith, who was part of the Lake Charles country-club set. When she lost her orthodontist to cancer, Shirley heard people at his funeral say his death was "God's plan." She felt an inner voice telling her that wasn't true, and that God wanted her to do something about the cancer deaths around her. She called Mike Tritico, who encouraged her to start her own organization to take advantage of her prominent place in Lake Charles society. He told her, "People think we are a bunch of hippies." Although she had never done anything like this before and was afraid of public speaking, Shirley started calling people she knew and was surprised when fifty showed up to a meeting at her home. They founded the Calcasieu League for Environmental Action Now (CLEAN) in 1982.

In a region dominated by industry, CLEAN helped to make environmental concerns more mainstream, with a membership of five hundred at its peak. CLEAN hosted public meetings about the health effects of toxins with the hope of educating industry people too. In addition to supporting the closure of the Willow Springs dump, CLEAN thwarted a proposal to burn toxic waste in the Gulf of Mexico, especially unpopular in Lake Charles, where Gulf winds could have blown the toxic ash.[98] Because of her social class, Shirley got a friendlier reception from politicians than Mabel, or for that matter, Mike.

WHILE SHIRLEY'S role was helpful, political access is not the same as political power. Peggy learned this when she watched a friendly politician vote against the dump closure just minutes after he told her he wouldn't. She grew up trusting public servants, but her political education began when she called the lead environmental regulator for the state and told him she didn't like toxic waste coming in to her region. "Well, lady, you might as well get used to it," he said. "He made the person that I became—this activist—because he was so *rude*," said Peggy. A former homecoming queen from East Texas, she was so upset that she was rude right back and then worried that he might sue her for "talking back." "That's how naive I was," she laughed, adding that the same regulator later became an industry lobbyist.

Naivete about the political system is one barrier that often keeps middle-class people like Peggy from joining working-class people, like Mabel and her mother, Beulah, who was known to "get down in it with those people," as Mabel put it. While questions of tone and approach can get in the way of cross-class organizing, research suggests that people from diverse social classes bring different strengths to social-change work, ultimately strengthening the movement. I heard such class differences as much as racial differences in the story of the dump campaign.

The people of Willow Springs, who relied on the land for sustenance, showed the greatest persistence. The working-class women of both races were described as fierce. Ruth and Beulah were willing to stand up to authority figures more directly than middle-class women like Peggy, who was better at calling people on the phone to invite them to a meeting. Peggy recruited a friend whose husband was an architect, so they used his copy machine. Mike, who had a science background, researched the chemicals in the dump and gave dramatic demonstrations of their risks to residents, regulators, and politicians. Herbert knew the land, where the frogs used to live and what other species were affected. Together, these people were far more powerful than any of them would have been alone—especially as their numbers grew.

In early 1985, they turned out six hundred people to a hearing on the dump-injection well. By then, they had shifted the spectrum of allies and convinced local politicians that the dump was dangerous. They nearly closed the facility entirely, but in the end part of the dump was kept open under much stricter guidelines.

"We had a rough time, but Michael and Peggy, they stuck by us," said Mabel. "And a lot more, they stuck by us."

NEAR THE end of our visit, Peggy said quietly that their kind of coalition couldn't happen today. When I asked why, she said that if Ruth Shepherd were still alive, she would probably be a Trump supporter. Peggy's comment reminded me that the United States became more politically polarized after the dump campaign ended. This has made it harder for people to ally based on a shared concern if they don't have a shared ideology or identity. Polarization has also made it harder for people to be neutral on any issue that has been politicized. To understand how this shift has affected the spectrum of allies, I reread a book set in Calcasieu Parish, which became a bestseller after the 2016 election of Donald Trump.

EMPATHY WALLS

Berkeley sociologist Arlie Russell Hochschild spent years visiting Calcasieu Parish, seeking to "scale the empathy wall" between left and right. In *Strangers in Their Own Land: Anger and Mourning on the American Right*, she explores why so many Louisianans who love the environment and hate pollution also hate government regulation of pollution. She describes this as "a keyhole issue," one that gives us a peak into a wider worldview. Some of the people she interviewed had been involved with CLEAN at its peak, including Lee Sherman, a retired pipe fitter who spent fifteen years working for Pittsburgh Plate Glass (PPG).

Years earlier, PPG repeatedly ordered Lee to dump extremely toxic "heavy bottoms" into an estuary southwest of Willow Springs. When this job ruined his health, he was fired without medical disability. Afterward, in the late 1980s, the government issued an overdue warning that local fish were unsafe to eat. Concerned about their livelihoods and mistrustful of government, fishermen were angry until Lee took the stage at a packed meeting. Uninvited, he dramatically confessed to his role in PPG's secret toxic dumping, convincing the fishermen the problem was real. Despite caring about pollution in his region, Lee told Hochschild that industry was overregulated, and climate change was "a bunch of hooey," an excuse for more big government. He even volunteered to distribute lawn signs for a candidate that wanted to cut funding for the EPA.[99]

Hochschild explores several theories about the rise of anti-government sentiment, including the loss of US manufacturing jobs due to globalization and the 2008 recession, which cost many people their homes. In the South, religious identity is especially strong, and the government is seen as anti-church. Some Southern whites are still bitter about their ancestors' harsh treatment by Union troops during the Civil War, and exploitation by Northern "carpetbaggers" afterward. Although most people receiving welfare checks are white, the Tea Party promoted the narrative that Black people are unfairly advantaged through government programs, tapping the old racist stereotype that they are lazy.[100] This narrative also fit what cognitive scientist George Lakoff calls the strict father worldview, where people succeed through competition, so advances by one group are assumed to be at the expense of others. This zero-sum game thinking helped fuel the backlash to the election of the first Black president in 2008, with the Tea Party bursting onto the political scene in 2009.

Another theory about the Tea Party's growth is that it was fostered by the Koch brothers. Founded on oil, Koch Industries was a major contributor to climate change and one of the top ten air and water polluters in the United States. In 1999 alone, Koch was found responsible for a deadly underground pipeline leak and for covering up the release of ninety-one tons of a known carcinogen at one of their refineries. Although they paid much less than the real cost of such accidents, government fines still cost them millions. To reduce environmental regulations, penalties, and taxes, the billionaire brothers quietly spent a fortune funding climate denial and a network of right-wing think tanks that manipulated racial divisions to protect their interests. The Tea Party appeared to be a spontaneous, grassroots movement, but many point out that it was based on ideas the Kochs had promoted for years. [101]

HISTORY SHOWS that elites have often tried to erect empathy walls between people who might otherwise join forces against them. During the 1640s, the Dutch colonizers of Mannahatta faced resistance from both enslaved Africans and the local Lenape. They granted some Africans "half-free" status and land to cultivate north of the Dutch settlement, where those Africans became a human shield in case of Lenape retaliation against Dutch violence. [102] A few decades later in colonial Virginia, Jamestown was burned down by rebels with diverse skin colors, indentured and enslaved alike. Although incited by a rich white man who wanted license to take more Indigenous land, Bacon's Rebellion terrified Virginia elites, who decided to prevent future interracial collaboration by solidifying the policies and ideas of race-based slavery. They passed a series of laws codifying social roles along racial lines and making slavery lifelong and intergenerational, while giving new status and protections to poor whites.

I first read about Bacon's Rebellion twenty years ago in *Learning to Be White: Money, Race, and God in America*, written by Thandeka, a Black Unitarian Universalist minister. Revisiting the book, I noticed that she describes the illusion of separation without using that exact phrase. Thandeka says, "Racial contempt would function as a wall between poor whites and blacks," protecting the enslavers and the wealth they amassed from *both* groups. This gave poor whites a "largely illusory" identification with the enslavers, which cost them psychologically, spiritually, and economically.[103] As a result, Irish minstrels in the nineteenth century sang about terrible urban living conditions and protested the gross disparity of wealth between rich and poor—all while wearing blackface, which scapegoated Black people, their competition on the labor market, and also their potential allies. My illusion that I was separate from this history was dispelled when a relative found a photo from the early 1940s of my teenage father performing a minstrel song in blackface with a cousin.

My ancestors were forced to leave Ireland due to the greed of English landlords, who exploited a deadly potato blight to increase their estates as 1.5 million Irish people starved to death. Those Irish who fled to American shores had a choice. They could ally with the people working beside them in the lowest-paid, most dangerous jobs; or they could vie for a better position for themselves and their children by siding with those above them on the social ladder. Most chose the latter. Irish Americans fought in US wars against Indigenous people, even as they sang songs about their own colonization in Ireland. They became so associated with policing that the anti-Irish slur "paddy wagon" is still used for law enforcement vehicles. Philadelphia still feels the legacy of Irish American-led trade unions that long excluded people of other races and ethnicities.

The most successful union organizing in US history occurred when workers were able to resist the divide and conquer game, improving working conditions and wages for all. It's no coincidence that the most vilified Irish American of her time was labor organizer Mother Jones, who in the late nineteenth and early twentieth centuries supported unionizing across racial lines, despite accepting the prejudices of her day. "The enemy seeks to conquer by dividing your ranks, by making distinctions between North and South, between American and foreign," she told a group of striking miners.[104] The violence used to suppress such organizing reminds us that European immigrants were incorporated into the US racial hierarchy, not only with carrots but also with sticks.

THE ROLE OF LABOR

Labor history reminds us that racial division was not inevitable. It was contested—and imposed. Once again, this is especially clear in Louisiana. Amid a violent backlash to Black political power after emancipation, Black sugarcane workers started organizing in the 1870s, eventually seeking support from the white-led Knights of Labor. The most powerful union in the country, the Knights helped to organize hundreds of sugarcane workers. During an 1887 strike, plantation owners brought in the militia, which massacred sixty people.[105] Only a few years later, Italian immigrants, recruited to work in sugarcane fields, allied with their Black coworkers. Historian Alan G. Gauthreaux cites this as the reason for several high-profile lynchings of Italians in the 1890s. He writes, "At the height of the white struggle to reclaim control of post-Reconstruction Louisiana, Italians challenged the nature of black and white racial bipolarity."[106]

During this same period, in 1892, twenty-five thousand workers of all races shut down New Orleans for four days. Several battalions of the state militia ended the general strike with the threat of violence. Afterward, New Orleans elites offered small wage increases and shorter hours to some workers, and union protection to none. A national economic crisis the following year helped their divide and conquer strategy. According to one account, "The solidarity across color lines displayed in 1892 was soon replaced by bitter hostility as wages plunged and many white dockworkers in New Orleans fought to deny African American workers access to the few good jobs available."[107]

CURIOUS WHAT race relations were like in Louisiana's oil and gas industry more than a century later, I asked a few different workers who described mixed progress. One older white man, who worked at the Exxon Baton Rouge facility, told me that when he started, almost four decades earlier, the few Black laborers were relegated to menial jobs. Then Exxon started to actively train and promote some of those workers. "All the older white mechanics out there resented that, big time," he said, recalling that shaking hands with a Black coworker could get you shunned by other white workers. Today, people from all backgrounds work at the Exxon complex, including women, though many in the highest-paying trades are still white men. Now Exxon would fire anyone who used the N-word, he said, but that doesn't mean there aren't a lot of guys who still have backward attitudes. He said that he was raised with those attitudes himself, but he started questioning what he'd been taught when he read a book about the Civil War. Having grown up poor, he realized that many poor whites died defending slavery, even though they never enslaved anyone themselves.

Hearing about the high cancer rates of petrochemical industry workers, I couldn't help but see a parallel. Many working-class whites died defending a system that didn't benefit them as much as it benefited those reaping the profit. Given that workers and communities close to industry both face dangerous levels of chemical exposure, they could be allies. This was exemplified by a pipeline that leaked in Calcasieu Parish for over a decade. Running between the Condea Vista and the Conoco refinery, the pipeline leaked as much as 47 million pounds of ethylene dichloride, a dangerous solvent linked to brain cancer and central nervous system damage. When the leak was discovered in 1994, the company deployed unprotected workers to clean up whatever toxins hadn't already disappeared into the soil and water. A class action suit was filed on behalf of people in Mossville. The workers filed a separate suit.

In 1997, Ray Reynolds, head of the local Oil, Chemical and Atomic Workers Union, reached out to Beth Zilbert, a thirty-five-year-old organizer from New Jersey that Greenpeace had sent to the area. "The workers at the plant are being treated just like that Black community. We're the frontline, just like they are," said Ray, who was slowly dying of toxic neuropathy, which he blamed on chemical exposure while working for Condea Vista.[108] "I want to stand shoulder to shoulder with the people in that community." Ray invited Beth over for his wife's world-class gumbo along with Debra Ramirez, who grew up in Mossville in the 1950s and '60s, when chemical accidents were so frequent that some people slept in their clothes to be ready for the middle-of-the-night, on-foot evacuations. Beth laughed recalling that Debra switched gumbo bowls when their hosts weren't looking to hide the fact that Beth was a vegan.

Despite Ray's hope for solidarity, Beth felt very alone when she went to meet the rank-and-file union members, most of whom were white like her. "No one in this room except Ray Reynolds wants me to walk out of here alive," she thought. It was one of the few union women who made her fellow workers listen. "Let's tell the truth," she said. "We know that this stuff is killers. She ain't lying. We all made a deal with the Devil when we came to work here. But when they came for my children, the deal was off." Debra and a few others from Mossville attended a Labor Day event with the workers, but lasting solidarity was never built.

THE MOST DIOXIN ON EARTH

About six miles south of Willow Springs, the small African American community of Mossville educated their own children during Jim Crow. It was known as the kind of place where people helped each other to raise a barn or dig a well. Folks fished, hunted, and kept backyard gardens. Some sold blackberries and mayhaws from roadside stands. Full of churches and one famous nightclub, Mossville included veterans, teachers, housekeepers, and if they could get the jobs, workers for the petrochemical companies that gradually took over the region. By the 1990s, fourteen such facilities surrounded Mossville. Two of the closest were Condea Vista, the company responsible for the pipeline spill, and PPG, the company that ordered Lee Sherman to illegally dump toxins in an estuary. Both corporations produced vinyl chloride, a toxic material used to make PVC pipes and plastic cutlery.

Debra Ramirez was one of many Mossville residents who spoke up about the impact of pollution, especially after she became a mother and her children got lesions on their scalps. Once, she called the local companies to inquire about suspicious smoke in the neighborhood, but each company blamed the other. Afterward, industry people asked to meet with her, and she brought a group of family and friends for moral support. "God had things pouring out of my mouth," she recalled. "I was telling them about the contamination of my air and drinking water. They were mad. I could tell by their eyes." Debra started going door-to-door in her community in the 1980s, raising awareness about the health effects of the plants.[109] Like Margie Richard in Norco, Debra recognized that the pollution traveled beyond her own community. "You cannot tell God's water where to flow or his air where to blow."

Mossville residents like Debra had been speaking up for many years when Greenpeace sent Beth Zilbert to Calcasieu Parish as part of its larger campaign against dioxin, which was released by both PPG and Condea Vista. "This was the place *on Earth*, not just in the US, where we found the most dioxin contamination in human beings," Beth told me. Going door-to-door in Mossville with Debra, Beth heard of large families that were suddenly having fertility problems, like endometriosis, and mothers concerned about the health of their children. "People were like, 'Come on in, sugar! You want something to eat? Are you thirsty?' They were the kindest people on Earth."

The white community in Lake Charles was less welcoming. After a few days, the local newspaper did a front-page smear story about Greenpeace descending on their community, which made Beth laugh since it was just her. No longer under the leadership of Shirley Goldsmith, some CLEAN members suggested Greenpeace should have reached out to *them*. One of the men in CLEAN said to Beth, "You're going door-to-door in *Mossville*? I don't even know where Mossville is." This was incredible to Beth given its proximity to Lake Charles. He implied that what she was doing was unsafe, but Beth knew the source of the real danger. One night a chemical release from Condea Vista gave her an asthma attack so bad she couldn't breathe and had to go to the emergency room. The people of Mossville looked out for her. "If ever there was motivation to say, 'This has to end,' Condea Vista gave it to me that night," Beth said.

KNOWN FOR their bold nonviolent direct action, Greenpeace was good at getting publicity and recruited local help. They hung a banner across the Westlake welcome sign, reading "No More Dioxin Factories, Stop PVC the Poison Plastic." They hung another banner from the state capital. They got Nickelodeon to film Beth as she did a program for children from Mossville and a nearby poor white community. When Beth asked how many children knew what brain cancer was, every hand went up. She asked how many knew someone with brain cancer, and every hand went up again. "I remember having to stop and gather myself," she recalled.

In the late 1990s, a series of studies reinforced what people in Mossville had been saying for years. The newly formed Louisiana Bucket Brigade trained Mossville residents to collect air samples, and Wilma Subra helped them interpret the data. They found vinyl chloride, benzene, and dichloroethane, all proven to be hazardous, well beyond legal limits. A test, by the Agency for Toxic Substances and Disease Registry, showed Mossville residents with dioxin levels at three times the national average. The government officials didn't believe these test results could be correct, but follow-up tests by Wilma Subra and others showed they were. A University of Texas symptom survey found 91 percent of respondents from Mossville had symptoms linked to chemical exposure, from runny eyes and bloody noses to cardiovascular, endocrine, and immune disorders.[110]

With Greenpeace's funding only temporary, Beth decided to stay in the area as a volunteer. As an organizer, she believed in working with anyone who was a potential ally, including the folks at CLEAN, despite their initial reaction to her arrival. She brought Debra to CLEAN meetings, and encouraged the group to bring people from Mossville onto CLEAN's board, but they did not. Debra focused her energy on the group she helped to found, Mossville Environmental Action Now (MEAN). Unbeknownst to any of them at the time, both groups were being closely watched.

THE SPY WHO EXPLOITED DIVISIONS

Mother Jones magazine uncovered the spy story. To undermine the lawsuits over the pipeline leak, Condea Vista hired a private security firm, Beckett Brown International (BBI), which was paid over $200,000 to steal privileged documents from the workers' lawyer and put local environmental activists under surveillance. BBI employees rifled through Beth's garbage, tracked the license plates of vehicles visiting her home, and took photos of Mike's mother's house.[III] They also hired an "operative" to infiltrate CLEAN. Based on the details revealed, everyone I spoke with believed the spy was Dick Rogers, a former algebra teacher. In his fifties at the time, Dick ingratiated himself with CLEAN leadership and was elected to the board along with his wife. From 1998 to 2000, he sent sixty-five reports to BBI, which encouraged him to learn as much as possible about the local organizations and their rivalries.[112]

In a 1999 report to its clients, BBI said that Greenpeace was calling CLEAN racist "because CLEAN refuses to focus their funds and efforts solely on the plight of the poor black communities and the environmental justice issue." The report implied that Greenpeace was racist for prioritizing people in Mossville. A follow-up report a month later noted that a potential donor had visited Lake Charles: "Two of his goals include mending the relationship between MEAN and CLEAN and stopping racist accusations."[113]

The BBI report reflected the growing national narrative among conservatives that any work to undo racism was actually "discriminatory." Those promoting this narrative even cited new research on DNA that proved that racial categories were illusions biologically. Instead of acknowledging that racist policies and ideas were social constructs that created real inequality, conservatives asserted that racism was also an illusion.[114] "You're the ones dividing people" became a frequent accusation against those who spoke out against racism, a talking point that was used against MEAN and their allies.

Everyone I spoke with about the spy story said that Dick exacerbated tensions between the groups, but they remembered different specifics. Mike claimed that Dick called a local store "N-Mart," a reference to "the N-word," prompting chuckles from other CLEAN members. Debra didn't remember that, but said she only met Dick a handful of times. "Dick didn't like me, and I felt it," she said. "His wife, she'd look under her eyeglasses. She looked like she didn't want to allow Blacks to vote." Lee Sherman, the former pipe-fitter who had dumped toxins, hosted a meeting with his wife, and she caught Dick accessing her computer. Beth recalled conflict over money as being the most explosive issue.

When CLEAN balked at putting Mossville residents on the board, a donor diverted a significant contribution from CLEAN to MEAN to pay for the buckets and the chemical tests. Beth said that CLEAN fell apart not long after. Focusing on what could be learned from the painful story, Beth noted, "They destroyed CLEAN, but not MEAN. The lesson from that is that you have to have a strong core. MEAN was formed by people who understood very clearly what their mission was and that it was literally about survival. They had no time for games." CLEAN, in contrast, had gotten more entrenched, which can happen when organizations have been around awhile. Another lesson was that people should not be naive about corporate espionage, though Beth warned that didn't mean you should be paranoid, just aware.

I heard yet another lesson in the story: It's hard to build a resilient cross-racial coalition when people don't have a common understanding of how racism works, especially how people's racial biases can aid the divide and conquer game. Union leader Ray Reynolds could see what he had in common with the people of Mossville, but many local whites couldn't, including many in CLEAN. "They blew us apart," said Mike. He was heartbroken that the parish environmental movement, which could turn out six hundred people in the '80s was depleted and fractured along racial lines by the late '90s. It seemed to me that, while Dick Rogers exploited their divisions, it was their illusion of separation that made them vulnerable.

SHORTLY AFTER this conflict, in 2001, Condea Vista was sold to Sasol, a South African company, which began planning an enormous petrochemical expansion that would displace most of Mossville. They won the support of local whites with the promise of jobs, even though many of the construction jobs went to workers brought in from other places. Described as the largest foreign direct-investment manufacturing project in US history, the massive complex was still not complete in 2018 when I first saw it on the ten-minute drive between Mabel's house and Willow Springs. "Oh my God, I didn't know it was this big," said Peggy. Several minutes passed as the General drove us past smokestacks, oil drums, and lots full of construction equipment. "I feel like crying," she said. "These were nice homes here."

I asked if people got compensated decently for their homes.

"The white people did," Peggy said bitterly.

There's still controversy about this. Sasol offered to buy out homes in Mossville for at least $100,000, plus 60 percent of the appraised value.[115] Some said this was a good deal for an area with low property values. Others argued that the offers were still not enough to replace what people were losing: paid-off houses, a close-knit community, and the land of their ancestors. Ultimately, most people in Mossville signed agreements with Sasol. Because the deal had a deadline, those who tried to hold out just got a gigantic polluting neighbor. The General told me that at one meeting a local official told people from Mossville that they got what they deserved if they didn't take the money.

"They are corrupt from top to bottom," the General said of the state government, which subsidized the $12.8 billion Sasol project through several different programs that gave huge tax breaks to industry.[116] Notably, this occurred in a region where many whites believe that government is unfairly helping Black people. Sasol's warm welcome was especially striking given that Greenpeace was labeled a "carpetbagger." When the fracture happened between CLEAN and MEAN, Beth said it was convenient for people to blame her, the outsider, another division industry exploits. Even activists from New Orleans were called "outsiders," while no such label was applied to multinational corporations like Sasol, Shell, or Formosa.

LESSONS FROM CALCASIEU PARISH

At the end of my first day in Calcasieu Parish, I turned on my hotel television for some distraction and stumbled on a movie about a mother dying of cancer. A few minutes into this story of love and loss, I burst out in a long cathartic cry for all the true stories I'd heard that day, and the long list of loved ones who had died. The indifference of industry and neighboring communities, the scale of human greed, the increased suffering that will come with increasing climate change—it was overwhelming. Now, when I have such feelings, I remember that Buddhist systems theorist Joanna Macy writes, "Our pain for the world releases us from the illusion of separation."[117]

One reason it's important to support frontline leaders like Mabel and Debra is that they are under no such illusions. As a result, they are often more committed than national or international organizations, whose priorities shift, often with their funding. Debra told me she felt abandoned by Greenpeace and sometimes resented their acting without consulting her community. In recent years, large groups like Greenpeace and Sierra Club have worked to face their own mixed histories and become better allies. Figuring out how outside organizations can be most helpful is important, since small communities on their own are rarely able to thwart a huge corporation, given all the pillars that hold up corporate interests.

Despite the unequal impacts, we all have a stake in stopping petrochemical expansion. Even Sasol, which is on the global list of companies most responsible for climate change, was hit by two hurricanes before construction was even finished on its Westlake facility. I left Calcasieu Parish more convinced than ever that we need to bring in more people, from across the spectrum of allies, to have any chance of stopping fossil fuel expansion. Later, I was inspired by an even more radical suggestion from author and activist adrienne maree brown who writes, "I want our infiltrators to be astounded into their own transformations, having failed to tear us apart."[118]

FROM HOME, I continued to follow developments in Louisiana. People seemed to be finding common ground in St. James

Parish, explicitly citing Mossville as a cautionary tale. "We don't want St. James to end up like Mossville," said Sharon Lavigne about the $9.4 billion plastics plant proposed for her community. The Formosa plant would add thousands of pounds of carcinogenic chemicals and an estimated 13 million tons of greenhouse gases per year, making it the largest new carbon emitter in the country.[119] A middle-aged special education teacher whose family had lived in the historic Black community of St. James for generations, Sharon prayed over what to do. She heard God answer, "Fight!" She predicted that St. James would not end up like Mossville because they had formed a coalition with other communities along Cancer Alley, which they renamed Death Alley. "We're not by ourselves," she said.

Anne Rolfes was also determined not to let what happened in Calcasieu Parish happen in St. James and said so explicitly on a Zoom panel with Sharon. Raised amid white wealth in Lafayette, Louisiana, Anne was taught to trust industry. She was grateful to women in Mossville who told her about the high cancer and endometriosis rates they faced. They helped to inspire her long career countering industry lies, including at the LMOGA conference I attended. I had the sense that Sharon and Anne were building the kind of grassroots support needed to resist a multibillion-dollar company. I remembered the words that had become my plumb line: "The crisis of the Earth is saving us from our illusion of separation." If that could be true in Louisiana, where the divisions are so deep and industry so strong, perhaps there is hope for all of us, I thought.

Still, I felt chilled by the shadow of the Sasol complex, and the boom in liquefied natural gas south of Calcasieu Parish during the years after my visit. Nationally, renewable energy was growing, but so were new fossil fuel projects. Looking at the organizations working to stop this, many seemed to be embroiled in internal conflicts—often about race. In some spaces, people seemed to spend more time criticizing those next to them on the spectrum of allies than they did the companies that profited from their division. As I grappled with these observations, I reached out to one of the smartest organizers I know to hear his thoughts about what could help us navigate race—given racism's role in dividing the spectrum of potential allies.

CHAPTER FIVE

NAVIGATING THE CURRENTS OF RACE

As a globally known trainer and strategist, Daniel Hunter has worked with a broad spectrum of organizations, including EQAT and 350.org, where he served as the director of international training. Daniel has also aided many groups going through racialized conflict internally, which is why I was dismayed when I once heard him say that no one in the climate movement had totally figured out how to navigate race. He'd quoted his mother—a white historian who was married to a Black preacher—who used to say that white and Black people kept missing each other through history. When Black people were willing to unite across racial lines, white people were not. Then, when there was a wave of white people ready to work in cross-racial solidarity, they found Black people tired of trying to collaborate and ready to go it alone. It was a sobering observation.

I was grateful that Daniel agreed to discuss this challenge on a windy spring day, sitting outside his old New Jersey farmhouse along the Delaware River. I shared my belief that we needed to bring in all kinds of people in order to make truly transformational change, but I didn't know how to do that given that ignoring racism and confronting it could both alienate potential allies. Tall, thin, and approaching his fortieth birthday at the time, Daniel nodded and observed that nothing around race is simple, even the categories people use to describe themselves. That said, he'd observed at least four different approaches to the dilemma, each with strengths and weaknesses.

"There are definitely good organizing approaches to get people in a room for a period of time to do a thing together that's race neutral," reflected Daniel, "but it's not stable." For example, an organizer might consciously, very consciously, he emphasized, look at a situation and say, "It's not going to be worth it for us to take on race. We need to fight this coal plant now."

Daniel and I had worked together on a race-neutral initiative, Choose Democracy, which he cofounded in the summer of 2020 to help prepare Americans for a potential coup attempt if Donald Trump lost the November presidential election. Although a successful coup would have been particularly dangerous for people of color, history showed that broad-based movements which included the political center had the best track record of defeating coups nonviolently. So while our trainers and other staff were racially diverse, our training content and media messaging emphasized that preventing a potential coup was in the interests of all Americans. In a matter of a few months, we got major media coverage and offered online nonviolence training to ten thousand people, using the pillars of power theory to identify what support Donald Trump would need to pull off a coup. After the January 6 insurrection failed to negate the election result, we disbanded until 2024, when Daniel relaunched Choose Democracy in a different form.[120]

Although avoiding the quagmire of race often seems expedient, Daniel noted that if you're trying to build a long-term coalition, sooner or later race will come up. If the issue has been buried, white people won't know how to handle it when it rears its head, often through allegations of racism within the group. "Once things explode, it's very hard to hold the group together, to carry the wounds," he said. Many white people don't know how to deal with the anger and frustration of people of color when it surfaces. Their defensiveness is part of what drains people of color. Daniel observed with a sad chuckle that there are many reasons movements fall apart, but race is usually one of them.

Daniel noted that the climate justice movement, and young organizers in particular, have rejected the race-neutral model. "They say, 'We have to deal with race.'" That means organizers today have to figure out how to talk about it. He asserted that there was a slightly better track record for groups that include discussion of race in their work, reading books together or having dialogue groups for their own growth, not just in response to a crisis. Then when tensions around race arise, people have some framing to hang on to. Of course, reading a few books is not enough to eradicate racism from our movements since it is so deeply ingrained. "It's just the water we swim in, so it's so hard to see," Daniel said, adding that he stopped leading anti-oppression trainings because they took too much out of him.

"Another approach is let's get race in your face, and let's get it often and early and directly and make it a thing that we're going to orient around," Daniel explained. He had sympathy for the years of frustration that led to this strategy and appreciated that it highlighted racial injustice. His concern was that primarily orienting around race could constrict the power of the climate movement rather than expand it at a time when massive bipartisan engagement is needed. He gave the example of his white Republican chiropractor, who is deeply concerned about climate change, having seen the impact of rising water temperatures on the rivers where he fishes. "A thermometer has no political leaning," he told Daniel, saying that he didn't understand the political divide over climate change. Daniel asked, "Do we have a way to have him in the movement, too?" Given the deep divides in our country, were there people who wouldn't join a group that welcomed the chiropractor?

THESE QUESTIONS were deeply connected to the question of what helps people to change. "I believe in truth telling—and, yes, our job as activists is to reshape people's hearts and minds. But if the bar to public discourse is perfection, we will have a very quiet crowd," Daniel asserted in an interview for the book *How We Win* by our mutual mentor George Lakey.[121] For that reason, Daniel felt that slamming those within the movement for their racism—or homophobia, or other prejudices—was counterproductive. It resulted in fewer people, not better people. While he believed that gently and deliberately encouraging people to shift their thinking was more effective than shaming, he also recognized that being overly patient or solicitous with whites who were clueless about racism drove away Black, Brown, and Indigenous people who were ready and eager to be involved. Another benefit of "race in your face," he told me, was that it pushes those who don't like this approach to at least wrestle with race and come up with other ways of addressing the issue. "That's extremely valuable."

"There's another approach that's been tried out a lot, which is come at it from a spiritual angle. This approach says, 'We're all one. We're all connected,' but it's infused with race awareness. 'We're all one, but we are not the same. We have different experiences.'" It was the approach of some of Daniel's heroes growing up, including Dr. King and Bishop Desmond Tutu in South Africa. Acknowledging that it was the approach he found himself most drawn to these days, he noted that how this message is received can depend on the messenger.

When Dr. King told a group of predominantly white college

graduates, "Whatever effects one directly affects all indirectly," people understood that he wasn't minimizing the pain of Black people. That's not necessarily assumed today if I, a white person, make the same point. Some white people try to show their concern about racism by talking only about the suffering of Black, Brown, and Indigenous people, but that can seem more rooted in pity than connection. This attitude is often labeled "white saviorism," especially when it leaves out the long and courageous history of people of color resisting injustice. My own experience as a speaker and activist is that acknowledging both our common stake and our unequal levels of risk helps me navigate between the currents of saviorism and denial of the role of race.

Near the end of our conversation, Daniel used the term "ecosystem" for a movement with different organizations playing different, complementary roles. I thought of how various types of flowers attract different bees and butterflies, bringing more pollinators to a diverse garden than a monochromatic one. It's a useful metaphor for groups that use the four different organizing approaches—which I'm calling race-neutral, educating about race, orienting around race, and emphasizing both oneness and difference. If we want to attract the maximum number of people to engage in any movement for change, we will need different types of groups to attract them, while being on the lookout for the pitfalls of each.

These models are a helpful starting point, but there are many further choices within them. The dominant race and generation of a group and the specifics of the issue will all impact what works to build love and power. To dig deeper into Daniel's four models, I found myself returning to stories I heard on the Gulf Coast as well as my own experience organizing in Philadelphia.

RACE-NEUTRAL ORGANIZING

General Honoré's website emphasizes that "all people have a right to clean air and water." In his public speeches, he's known to ask, "How many people like clean air?" while he raises his own hand with a sly smile. The audience follows suit. "How many like clean, safe water? How many people like their crawfish without any oil on them?" The word *oil* stretched out by his Louisiana drawl. People chuckle, their hands in the air.[122] His questions and their obvious answers build common ground.

After his retirement from the army, the General wanted to do something useful beyond book writing and public speaking. "Just about that time, I got a call from some Cajuns down on the bayou," he told me. Reckless drilling by a Texas-based company had caused a catastrophic sinkhole on the western outskirts of Cancer Alley in the predominantly white community of Bayou Corne. Over thirty acres of land were gradually swallowed up, towering cypress trees dramatically sucked into the Earth. Gas spread deep underground, polluting the aquifer and creating an explosion hazard. Over three hundred people were displaced. Struggling to get their concerns addressed, they reached out to the General, the "Black John Wayne," heralded for breaking the government logjam and evacuating New Orleans after Katrina eight years earlier. He used his celebrity to arrange for them to come up to Baton Rouge to speak to the press and "put some heat under the capitol," as he put it.

The General's military rank and reputation as a strong leader helped gain the trust of Tea Party members in Bayou Corne. Working along with more experienced advocates, he told me they got a bill passed saying that if people were forced to move from their homes, industry had to pay them replacement value and not market value. "Because if they pollute where your house is, nobody wants to buy it," the General explained. "So, if you had any value it's gone."[123] That experience turned the General into a passionate advocate for communities suffering from pollution, regardless of their race or politics. It struck me that this bill, while race-neutral and passed because of the suffering of a white community, would help many Black and Brown communities, who for the reasons I saw in Norco, had an especially hard time getting a fair price for their homes.

As Daniel mentioned, race-neutral strategies have fallen out of favor, especially with young activists and those who come out of the environmental justice and climate justice traditions. I suspect this was the context of the campaign against the Entergy gas plant in New Orleans East. At a small New Orleans bar, Pat Bryant—a Black activist with many decades of experience—told me that during that campaign he had argued for more of a focus on the shared economic cost of the plant. He believed that the threat of increasing utility rates would bring in more white people than talk of environmental justice. Pat lost the argument. Cost did not become a unifying argument to prevent the project, but it did become a focus of shared resentment after the plant was built, especially when Entergy failed to resume power service quickly after Hurricane Ida, leading to increased heat deaths from lack of air-conditioning.[124]

THE GENERAL and Pat Bryant understand racial inequality and are not afraid to name it, but they choose for strategic reasons to emphasize our common stake. That's very different than avoiding race without appreciating its role as a pillar of power. "I can't say it hard enough. We don't have time to argue about social justice," argued Jonathan Logan, a middle-aged white man, who helped form the break-off group Extinction Rebellion (XR) America after a schism developed among the leadership of XR US. Some wanted racial justice named as a key part of the economic transition they demanded, and some, like Logan, thought this would be unnecessarily polarizing. "If we don't solve climate change, Black lives don't matter," he quipped, offending those who believed that Black lives mattered now.[125] In another interview, Logan proclaimed, "We are one planet, with one humanity, and one common future—kind of like Star Trek world."[126]

Logan's flip Star Trek reference, while apt on one level, lacked sensitivity to those trying to survive in this world, not a futuristic one. It dismissed the huge racial disparities of who is most likely to die from climate change globally and who is most likely to experience police violence during civil disobedience. While advocating oneness, Logan and other leaders missed a key expression of true oneness: compassion. They also missed an important lesson of interconnection—that our problems are also interconnected. Although purportedly intended to build a more united movement, calls for race neutrality split the group and drove some Black, Brown, and Indigenous people out of XR altogether.

The pros and cons of race-neutral organizing are especially clear in the history of the Calcasieu Parish environmental movement. Like my first bumper sticker, "We All Live Downstream," Herbert Rigmaiden and Mike Tritico emphasized the common risk the Willow Springs dump posed without even mentioning the obvious fact that Black people bore the harshest health and economic impacts. They could turn out six hundred people to a hearing in 1985, but they collapsed in the '90s amid allegations of racism. "The people of Mossville, some of them still think that we double-crossed them, but we didn't," Mike told me vehemently. Their model of collaboration had worked, he insisted, but once allegations of racism were made, everybody pulled away. "It was a terrible, terrible thing. They blew us apart," he concluded.

Mike went on to say that there wasn't much racism when he was growing up in Louisiana, an incredible assertion about the Jim Crow South. "Everyone got along," he insisted, sharing examples of people being friendly to each other. Instead of understanding racism as a collection of policies and ideas that produce and justify racial inequality, Mike thought of racism as overt animosity toward people of different races. His old friend Peggy Frankland gently challenged his perception and pointed out that racism had always been there; it just wasn't discussed openly when they were young (at least in their communities, I thought). Although both were Southern whites in their seventies, Mike and Peggy were clearly starting from different but unnamed definitions of *racism*. This is part of what Daniel meant by people not having "handles" for conversations about race.

Although Mike was justified in blaming the industry spy for blowing them apart, it seemed to me that whites in Calcasieu Parish had given Dick the ammunition through their own unwillingness to acknowledge racism. It was hard to imagine CLEAN educating its members about race, but maybe it would have helped.

EDUCATING ABOUT RACE

How a group educates about race will differ depending on the demographics of a group. The coalition against the gas plant in New Orleans East made space for Black, white, Asian, and Latino people to listen to each other's stories. For predominantly white groups, it's important that what is often a small number of people of color aren't expected to become spokespeople for all people of color or carry all the emotional labor of this work. It's also important that the work is woven into the overall mission, not irrelevant, or worse, performative—actions that seem to be motivated by a desire to look good rather than a real commitment to change.

Educating about race developed gradually within EQAT, intertwined with aspects of other models. Our Bank Like Appalachia Matters! campaign was race-neutral both in its demands and organizing strategy. Still, there was much for a middle-class Philadelphia-based group to learn about the struggles of working people in Appalachia. Many of us became aware of our physical and emotional distance from the devastation that Appalachians felt acutely. This dynamic became even more painful when we took on a campaign that highlighted racial inequality in our own region.

Many EQAT members were forced to acknowledge their separation from Black and Brown communities only a few miles away from their own. We tried to make space for people's feelings, knowing that shame can prevent people from taking action. Aware that some organizations try to make themselves look more diverse than they really are, we tried not to tokenize our few members of color by overusing their photographs on our website or disproportionately pushing them into public-speaking roles. At one point, Rev. Rhetta Morgan, whom I had introduced to EQAT, commented, "You're trying so hard not to tokenize me, I'm not sure you really want me." I learned to explicitly ask people how they want to be treated and what roles they want to play. Rhetta, a Black interfaith minister, later joined the EQAT board and brought a strong spiritual frame to the work of confronting racism as we worked to do that more explicitly as an organization.

The Power Local Green Jobs campaign aimed to address climate change while creating economic opportunity for the neighborhoods that needed it most, so we offered participatory trainings to help our members learn about the impacts of poverty and racism in our own region. Our partnership with POWER Interfaith was particularly helpful, as well as relationships we built with Black Philadelphians advocating for solar. EQAT members were moved when we held nonviolent direct actions at our utility's headquarters, and speakers shared their lived experience of high asthma rates and high unemployment in their communities. We also did trainings designed to help us be more aware of our own biases and the unspoken norms of the group, which can inadvertently exclude people of different backgrounds. There was broad support for this work when it was framed as part of campaign building, but generational and ideological differences still emerged.

We agreed that shaming people for mistakes, or "calling out," would not build the campaign. Inviting people into dialogue, or "calling in," was more likely to teach them without alienating them. Still, staff of color sometimes thought that white leaders like myself responded too gently, or too privately, when white members made comments that revealed racial bias. One such incident occurred when we walked one hundred miles around PECO's service territory. In a predominantly Black Philadelphia neighborhood, a white EQAT member commented on the poverty she witnessed in a way that felt judgmental to a staff member of color who overheard it. Speaking to the volunteer afterward clarified for me a major challenge of educating about race. We wanted our members to learn about the disparities between communities, but in a way that built connection and common ground rather than reinforcing separation or superiority. Learning to have empathy, rather than pity, required real relationship work, not just walking through a neighborhood— or reading a book.

SEVERAL MONTHS into the campaign, our electric utility invited EQAT to an all-day "Solar Stakeholder Collaborative." We were glad that PECO felt the need to respond to our campaign, but suspected that it was just a PR ploy, so our strategy team decided to send representatives from EQAT and POWER but not promote the event more broadly. The day before the meeting, I learned from a mutual friend that John Bowie—a Black man who had been advocating for solar in his North Philly neighborhood since before our campaign—was upset to be left out. Wanting to really listen to John, I called my friend Ingrid Lakey to admit and work through my defensiveness before calling John. From him, I heard how EQAT deciding what was or wasn't a significant meeting fit the pattern of a predominantly white and middle-class group accessing power holders, while frontline stakeholders were excluded from "the table" during conversations about their own communities.

At John's request, I advocated for him to be included in the meeting, where he spoke about the lack of economic opportunity in his neighborhood during a breakout group that included PECO executives. Unlike those who described poverty from a place of disconnection, John was describing the community he loved.

As some of the founders rotated off the board and the average age of our board dropped, the time we spent talking about race increased, especially during the COVID-19 pandemic, when organizing actions became more difficult. A small group put together a curriculum of readings and podcasts and organized small groups to discuss them over Zoom. I noticed that instead of pulling away, which sometimes happens when white people are pushed to confront racism, many of our members seemed to grow in their understanding of and commitment to racial justice. I also noticed that educating about race posed a time and emotional burden for our Black and Brown members, who, while few, were all in leadership positions. As Daniel noted, working with white people could be draining. There were a few times when leaders of color pulled away emotionally to protect themselves. I suspected that orienting around race emerged in response to this dynamic.

ORIENTING AROUND RACE

The HBCU Fighting for Our Lives Conference was grounded in love for the people the academics/advocates aimed to help protect. Unlike other academic conferences I've attended, there was room for spirituality and emotion. It also oriented around race in a way that enabled Black students to claim their place in a long tradition of struggle against racism. I could feel the importance of that lineage as the audience listened to pioneers like Robert Bullard, Vernice Miller-Travis, and Beverly Wright, who changed the narrative in the 1980s and '90s by proving racial disparities in pollution exposure. Just as the HBCUs were established to provide Black students opportunities denied them by predominantly white institutions, conference organizer DSCEJ was founded in 1992 to support communities like Norco and Mossville, which had not been represented in national environmental organizations.

Watching the toll that working with white people took on EQAT leaders of color gave me a greater appreciation for how welcome identity-specific spaces could be.

Sitting at her dining room table in New Orleans, Jayeesha Dutta told me about an initiative she was part of with five other women who identified as Brown. They came from Indigenous, Latina, and Asian backgrounds, but they shared an experience of being racially marginalized. Based across the Gulf of Mexico from Texas to Florida, they named their project "Another Gulf Is Possible" with the goal of supporting each other's work for a just transition away from fossil fuels toward a new kind of energy economy.

Born to Bengali immigrant parents, Jayeesha described herself as an artist and activist who works on creating new narratives, including through film. She told me that she has experienced racism, not just in the Gulf Coast where she was born but in New York where she grew up and California where she used to work. "It's so far under the rug," she said of racism in California and New York, "but it's the same shit, to use a curse word. It's the same shit everywhere." In Louisiana, she said, it was "in your face every day where you smell it," which in some ways made it easier to deal with than spaces where racism was hidden and denied. She asserted that we would never get to real solutions until we deal with the roots of racism and all the original sins that created this country—the stealing of land, stealing of labor, and stealing of lives.

Jayeesha pointed out that the racial discourse, especially in the South, has been framed as Black/white, when in fact there have been Bengali Americans in New Orleans since before Jim Crow. "Obviously, Indigenous folks and Latinx folks have been here all along. Centering our voices very intentionally breaks up that Black/white binary." Making our stories around race more complex was inherently challenging of the dominant narratives. Being based in the South—the US region with the largest Black population—meant that Jayeesha felt part of the lineage of groundbreaking activism of the past, as well as a culture that moves more deliberately and is more focused on community. "I remember back in Oakland, we'd just have a bag of chips in the meeting." In the South, she said, people make time and space for cooking and sharing food, which are crucial in bringing people together. She predicted that the Gulf Coast, so steeped in the fossil fuel economy, would play a groundbreaking role in finding another way forward.

ON A visit to Houston, I met another member of Another Gulf Is Possible, Yudith Nieto, whose family immigrated from Mexico to Texas when she was seven. Because it was less expensive, they came to the small community of Manchester, which was embedded in Houston's enormous chemical corridor. On my way to meet her at the community center, I noticed one Manchester yard brimming with bright flowers so close to a chemical drum that a child could throw a ball from the yard and hit it. A frequent spokesperson for her community, Yudith described the same smells, explosions, and health impacts common in Cancer Alley. As immigrants, they also faced the intimidating presence of Homeland Security and police. For Yudith, all her identities inform her work, including being young and queer. She noted that her family had lost touch with their Indigenous roots because of colonialism, but she included practices that honored the land in her work.

Part of what Yudith appreciated about being part of a collaboration of Brown women was space for creativity, intuition, and healing. They could share their emotions and not worry that they would be stereotyped as angry Brown women "and just kind of waved away as, 'You're too hard to deal with,'" which often happened in predominantly white or male spaces. When Brown women are angry or feeling passionate, she said, "It's because we see the urgency of finding solutions to what's going on." When they had space to explore solutions themselves, they were rooted in what was needed rather than abstract ideas imposed from the outside.

IDENTITY POLITICS

Hearing these stories reminded me of the Combahee River Collective (CRC), a group of Black women who began meeting in 1974 to share and sharpen their political analysis. Collectively they had been engaged in different aspects of the freedom movement, the women's liberation movement, the anti-war movement, and what was then called the gay liberation movement. The women of CRC saw male leaders of Black organizations replicate patriarchal patterns and white leaders of female organizations replicate racist ones. Some of them also experienced oppression around their social class and sexual orientation. Recognizing these systems of oppression as interconnected, they wanted a space of their own to explore comprehensive solutions.

"The most profound and potentially most radical politics come directly out of our own identity," the CRC asserted in a 1977 statement, which introduced the term "identity politics." They explained that their politics evolved "from a healthy love of ourselves, our sisters and our community."[127] They saw coalition building as key to winning all their struggles, and assumed that people would join coalitions out of their own motivations—like the union leader in Calcasieu Parish who realized that he was exposed to the same toxic chemicals as people in Mossville.

Activist-scholar Keeanga-Yamahtta Taylor writes that for the CRC, "Solidarity did not mean subsuming your struggles to help someone else. It was intended to strengthen the political commitment from other groups by getting them to recognize how the different struggles were related to each other and connected under capitalism." She explains, "The women of CRC did not define 'identity politics' as exclusionary, whereby only those experiencing a particular oppression could fight against it. Nor did they envision identity politics as a tool to claim the mantle of 'most oppressed.'" Their ultimate goal was liberation—for themselves and for everyone in the world.[128]

This analysis was threatening to power holders. The very idea that people who have been marginalized could claim strength and insight from their marginalized identities was a powerful subversion of racist, sexist, and other discriminatory ideas. That was the context of the CRC statement, as Black, Latino, Indigenous, female, and queer people built powerful movements in the 1970s that would have seemed impossible two decades earlier. In hindsight, it's not surprising that the concept of identity politics was attacked as part of a concerted backlash to progress. The term was repeatedly belittled, its meaning twisted to imply division or special interests. Today, any oppressed person who claims strength and insight from a marginalized identity while calling for justice may be accused of playing identity politics, rather than asserting a politics grounded in their own experience.

THE BACKLASH against identity politics tapped into a new identity insecurity among some white people, especially conservative men. For those who were also working class, their fears were exacerbated by economic changes, such as the decline of coal mining in Appalachia and manufacturing jobs that moved overseas. Eventually, the shame and isolation of workers who felt left behind were harnessed to justify the dismantling of diversity programs, as well as whole government departments, even those that were creating new solar and wind jobs across the United States.

Years before Donald Trump's 2024 reelection, some racial-justice advocates began grappling with the ways that our narratives have backfired, often with the help of right-wing distortion. After years of policy advocacy grounded in love for her own Black community, Heather McGhee realized that people who wanted clean air, good schools, fair wages, and a functioning safety net needed a more effective way to frame these issues. In *The Sum of Us: What Racism Costs Everyone and How We Can Prosper Together*, she explores the shared but unequal cost of racial division and shows there is a "solidarity dividend" when we explicitly fight for everyone.[129]

Some activists on the left have inadvertently aided the right-wing divide and conquer project by essentializing identity in ways that make finding common ground more difficult. When EQAT began educating about race, one recent graduate from a progressive college said she had gotten the message that, as a white person, she was wrong if she wanted to try to address racism, and she was wrong if she didn't, as if her identity doomed her to inevitable wrongness. Similarly, I was once lectured by a white climate activist who, in his seventies, was getting to know a few Black people for the first time in his life. He confessed how ignorant he had been about racism. Then, he warned me that I should be wary of speaking to our Black partners at POWER because, as a white person, I might inadvertently harm them. Never mind that POWER was founded by skilled and resilient organizers, who in the face of systemic racism chose to build grassroots power by mobilizing people across lines of race, class, and religion.

It's true that white people can do harm. Most of us have, including me, and it's important to be mindful of our impact. The problem is that this Yale-educated lawyer had learned about racism in a way that portrayed Black people only as fragile victims—instead of people with agency who could choose what kind of organizing they wanted to do. He had also gotten the message that there was nothing he could do but keep his mouth shut around them. While that was probably a wise first step for a man with his social conditioning, he had not actually overcome his assumption that he knew better than everyone else. So, he lectured me, a woman. I felt tempted to call him out, but because he said he was attempting to learn, I spoke more gently. I invited him to notice that he had not even considered that I might have a much longer history of interracial experience than him. He apologized, but still didn't show any curiosity about my perspective.

Essentializing identity has another downside. I've seen corporations try to deflect allegations of environmental racism by hiring people of color to represent them. The most uncomfortable example was a contentious hearing in St. James Parish. Formosa was trying to build a plastics plant in a 91 percent Black district, so the company sent a Black woman to refute the community's concerns. She sat stone-faced next to Jim Harris, the white PR specialist who denied the existence of environmental racism at LMOGA. The campaign eventually thwarted Formosa's plans by using Daniel's fourth strategy, centering oneness and difference.

CENTERING ONENESS AND DIFFERENCE

"I know you can't sleep at night because you live in St. James, too," Sharon Lavigne said to the parish officials who had approved the Formosa plastics plant. She asked them to rescind their approval by appealing to their common interest as neighbors in the same parish. "When I am poisoned, you will be poisoned, too," she said. Later, when I spoke with Sharon at a national mobilization in DC, I asked her why people didn't seem to understand that they were poisoning themselves and their own families. "They're taking money from the company, and they don't want to see it," she said firmly.

Although the parish council was majority white, Sharon added, "It doesn't matter what color they are because they all vote for it. They don't realize they're not just hurting us. They're hurting themselves, too." I noticed that Sharon called for cross-racial solidarity while blaming the greedy few, a point she reiterated when I asked if racial divisions had been a barrier to her organizing. Shaking her head, she said that the white people whom she took issue with were the wealthy ones who moved out. "They knew years prior to all of this that Formosa was coming in, but they didn't let us know. They grew up with us, they went to school with us, and then they kept it a secret from us." She felt that was selfish and gave her community a late start in organizing against the plant. In contrast, she described the whites who remained as "family." They were supportive of Sharon and the campaign.

By the time of our October 2021 conversation, the Formosa plant had been stalled after a few years of continuous grassroots opposition. I asked how they did it. "People power," Sharon responded. She founded RISE St. James with ten people meeting in her living room. By the next month, it was twenty. Sharon and her daughter, Shamell, organized small marches, which energized people and built Sharon's confidence using a bullhorn. They formed a coalition with other Cancer Alley communities and benefited from the support of advocates like Anne Rolfes, who had over twenty years of experience in such fights. "The more organizations we got together, the more we build up our power," recalled Sharon. "You can do anything if you got people together."

When I asked Anne about their success, she began by acknowledging Sharon's role as a strong community leader who spoke about the campaign at every opportunity, reaching out to local, national, and even international media. The Formosa campaign also benefited from other earlier campaigns, she noted. There were Louisianans who had joined the Indigenous-led resistance at Standing Rock and came back inspired to stand up to the Bayou Bridge Pipeline. Although they didn't win that campaign, they built relationships with people in St. James, where the pipeline terminated, and then supported them when they were targeted by Formosa. "There haven't been many campaigns along Cancer Alley where you have so many people involved," Anne observed.

AT A 2018 permit hearing, several people I had interviewed about other struggles spoke in opposition to the Formosa plant. Cherri Foytlin was fresh from the pipeline fight, where she had been brutally arrested. When her mic was shut off, allegedly due to her use of the word *shit*, she yelled that the officials at the hearing were more concerned about profanity than poisoning people. Pat Bryant reinforced this point. "What you all are doing is profane," he told representatives of the Army Corps of Engineers, the Department of Environmental Quality, and the Louisiana Department of Natural Resources. During the three-hour hearing, Mark Nguyen of VAYLA talked about all the fish killed by the Formosa accident in Vietnam, which we had learned about together at our lunch at a Vietnamese restaurant in New Orleans East.

Knowing that testifying at hearings was not enough to overcome the power of industry, RISE St. James also held many creative actions. Sharon, who is a devout Catholic, often used religious language and imagery. In one action, the group reenacted the biblical story of Jericho, whose walls crumbled after Joshua circled them six times. They imitated the prophet by walking in a circle, and on their seventh rotation raised their arms to the heavens and appealed to God to stop Formosa.[130]

There were several preachers involved in the campaign, including Rev. Manning, who was thrown to the ground when arrested during a Coalition Against Death Alley event. In 2019, Manning founded the Greater New Orleans Interfaith Climate Coalition, bringing together Jews, Muslims, Christians, Bahá'ís, Indigenous people, and others around care of the Earth and climate. "The one thing that unites us is love," he professed on a Zoom panel with Sharon and Anne. Centering both oneness and difference, he said, "We cannot stand to be divided any longer," then acknowledged that the land along the Mississippi River was "bought and paid for by the blood, sweat, and tears of former slaves," people whose souls were still crying out for justice.

The plantation history of St. James was highlighted by the controversy over unmarked graves on the property where Formosa planned to build. Since unmarked graves almost certainly belonged to formerly enslaved people, RISE St. James held a Juneteenth commemoration at the site in 2020, blessing the ground with holy water and laying flowers. They argued that Formosa's permit should be rescinded since they hadn't disclosed the cemeteries on their application, even though the company knew about them. Formosa tried to stop people from visiting the graves, but a judge ruled that if Formosa wanted to be a "good neighbor," as it claimed, the company should be sensitive to the need for healing and honor the sacredness of burial sites. Listeners teared up during the judge's speech, and RISE St. James members proclaimed it as God at work.[131]

Under continuous pressure, Formosa announced a delay in construction, citing the COVID-19 pandemic. Then the Army Corps of Engineers suspended the original permit due to errors and in 2021 announced that it was commissioning a full environmental impact statement with special attention to environmental justice. This was not a definitive win, but Sharon felt confident. "They're not going to build two miles from my home. We're going to fight till the end." She added that she wasn't just concerned about her own neighborhood. If Formosa tried to build in neighboring St. John the Baptist Parish—another beleaguered part of Death Alley—she would fight alongside them. "All of us, we're all in this big pot of gumbo together. We've got all different ingredients in that pot of gumbo, so we're going to work together, and we're going to help St. John if that's where Formosa thinks they're going to build."

LESSONS ON BUILDING COMMON GROUND ACROSS RACE

"We actually know what wins," asserts Anat Shenker-Osorio, a widely sought expert on how progressives can make their language more effective. At least in elections, where getting the majority is necessary, she says completely race-neutral messaging falls flat because it fails to directly counter race-baiting as a divide and conquer tactic. On a range of progressive issues, the common Democratic tactic of saying, "This will be good for everyone, especially Black people" is also less than ideal. That approach, she says, makes Black people feel like an afterthought, while reinforcing the division in white people's minds. Inadvertently, this makes whites more susceptible to "dog whistle politics," the coded racial messages the right wing uses to undermine faith in government, not to mention collective action.

What works with the broadest range of people is messaging that says, "We all need this, whether we're white, Black, or Brown." Shenker-Osorio's research also shows that it helps to acknowledge three other points: the greed of the few; racial scapegoating as a divide and conquer tactic that hurts everyone; and racial solidarity as a path to shared prosperity.[132] The campaign against Formosa incorporated all of these strategies.

In Louisiana, I'd met several contemporary organizers who helped people to find common ground, not by ignoring racial or class disparities but by highlighting other shared identities. For some, it was shared place, like being from St. James for Sharon or being a Louisianan for the General. For Rev. Manning's group, it was faith, even if the faith traditions were different. Globally, young people were forging a shared identity based on their fear of climate catastrophe and anger at the failure of their elders. In each of these cases, a shared identity connected to a shared value: breathing clean air, eating clean crawfish, living morally, or living at all. Shenker-Osorio advocates appealing to such shared values to build common ground.

AT FIRST, I conflated shared identities and values with the spiritual framing of oneness, but Matthew Armstead pointed out that they were not quite the same. Once an organizer with EQAT, they were working with We Make the Future, an organization founded in 2021 to apply the research of Anat Shenker-Osorio and others to help progressives communicate more effectively. Matthew explained that spiritual oneness was deeper and more universal than mutual interest, lasting even when group interests diverged. I thought of the Quaker concept that there is "that of God in every person," and how I'd struggled to see it in the oil and gas executives at the LMOGA conference.

In the years since that experience, I'd become increasingly convinced that we need to challenge power holders doing harm, even as we recognize that we are all connected. This feels especially important at a time of political polarization, where demonization has become the norm and political violence is on the rise. Treating people who disagree with us with love can be hard, but treating them with hate or condescension pretty much guarantees that they will never join us, even if they come to recognize their own stake in a more just and sustainable way of doing things. In this way, the oneness and difference approach is both spiritual and strategic.

This realization made me even more intrigued by Indigenous-led campaigns that courageously challenged those threatening the web of life while still calling all people relatives. The fact that Indigenous cultures also recognize other species as relatives further expanded my understanding of the message that the crisis of the Earth is saving us from our illusion of separation.

PART III

The Great Turning

CHAPTER SIX

ALL OUR RELATIONS

As sunlight softened over the gentle headwaters of the Mississippi River, thirty-four-year-old Gina Peltier stood on a narrow metal bridge with two other Indigenous women and four white allies, singing an Ojibwe prayer for the endangered water. Nearby, noisy crews from the Enbridge corporation had blasted under the riverbed, spilling undisclosed chemicals. On this August evening in 2021, the workers were preparing to thrust the Line 3 tar sands oil pipeline under the sparkling water. When the Clearwater County Sheriff's department arrived, the young deputy did not order the Enbridge workers to stop even though the company had violated numerous regulations in its rush to get the pipeline installed. Instead, he ordered the seven people praying on the bridge to leave. When they refused, they were arrested for being a "public nuisance."

A citizen of the Turtle Mountain Band of Chippewa, Gina pulled out her phone for a quick Facebook Live while she waited for the sheriff's van to arrive. Acknowledging the fish who were suffering from the recent drilling fluid spills, she took off her glasses to show her beautiful round face, framed by long black braids—not out of vanity, but as evidence in case anything happened to her overnight in jail. She asked her audience to excuse her shaking hands. "I've never been to jail before," she confessed with a slight smile. "I'm doing this for all you guys. I love you all."

A FEW weeks earlier, north of the Mississippi headwaters, Gina and I sat down along the grassy banks of the Red Lake River, where we were both camping. Volunteer allies had vetted me, and after a Zoom training and a phone call, they'd assigned me to one of several camps along the pipeline's route. I'd come to both support and learn from the Stop Line 3 campaign. Amid a soft breeze, Gina shared the Indigenous teachings that motivated her. "We're taught to respect everything," she began. "Everything has a spirit. The wind has a spirit. Water has a spirit. Grass has a spirit." Even as you scoop a spider out of your tent, you respect it as a cousin. Because we are all relations, she explained, "Indigenous people feel a special obligation to defend Mother Earth, defend the sacred, defend the water."

In Minnesota, known as "the land of ten thousand lakes," they were also protecting *manoomin*, the musky wild rice that grows on fresh water and is important to Ojibwe history, spirituality, and sustenance. Gina explained that Line 3 would be built across land that Ojibwe nations relinquished in treaties with the US government. Under those treaties, the Ojibwe were promised continued rights to fish, hunt, harvest rice, and perform spiritual ceremonies on treaty land. Leaks that polluted the water would threaten all those activities.

At the time, Gina was on staff with Honor the Earth, an Indigenous-led organization that was part of the coalition to stop the pipeline. Led by Ojibwe women of different nations, the seven-year campaign had galvanized thousands of Minnesotans to testify during the permitting process. When the pipeline permit was approved anyway, they tried to stop it by putting their bodies in the way. From different Indigenous nations and from non-Indigenous communities across the United States, thousands of people had joined the Stop Line 3 campaign. Many spent days, weeks, or even months in camps dispersed along the pipeline route. Gina and I met at the Red Lake Treaty Camp.

The green grass of the camp stretched downhill to the river, where we fetched dishwater and cooled off in the gentle current. Next door, a chain-link fence bordered the compound of packed dirt occupied by Enbridge, where noisy machines were preparing to drive a section of the pipeline under the river. The scene reminded me of the Anishinaabe Seventh Fire Prophecy, which says that humanity will have to choose between two paths. "One of the roads is soft and green with new grass. You could walk barefoot there," writes Robin Wall Kimmerer in her book *Braiding Sweetgrass: Indigenous Wisdom, Scientific Knowledge, and the Teachings of Plants.* The other path is scorched and hard. "If the people choose the grassy path, then life will be sustained. But if they choose the cinder path, the damage they have wrought upon the earth will turn against them and bring suffering and death to earth's people."

A professional botanist and member of the Citizen Potawatomi Nation, Kimmerer describes the green path as "lined with all the world's people, in all the colors of the medicine wheel." These people understand the choice ahead and "share a vision of respect and reciprocity, of fellowship with the more-than-human world."[133]

WHILE SOME allies came to support Indigenous rights, Gina noted that others understood that they also had a stake in protecting Mother Earth. "Pipelines leak," she said. "This water will flow south and endanger not just Minnesotans, or not just an Indigenous culture. It's millions and millions of people that it's going to affect." Crossing hundreds of bodies of fresh water, the Line 3 pipeline would be especially prone to leaks because of the corrosiveness of tar sands oil and the high pressure needed to move its weight. Not only do tar sands contain benzene and other carcinogens, but the sticky material is also arduous to clean up.[134] Even if the pipeline never spills, burning tar sands oil is terrible for climate change. While those most immediately endangered by the pipeline were in northern Minnesota, the people and animals ultimately affected by Line 3 spanned the globe.[135]

Visiting the Mississippi headwaters gave me a better sense of how this waterway connects the fates of diverse people and species. Known in Ojibwe as *Misi-ziibi* (Great River), it provides drinking water to almost 20 million people in fifty cities on its 2,300-mile journey from northern Minnesota to the Gulf of Mexico. Along the way, tributaries flow into it, as well as pesticides from Midwestern farms, sewage, pollution from factories, and toxins from Cancer Alley. The Mississippi is also home to 260 species of fish and 145 species of amphibians and reptiles. It's a major migratory path for birds, which is how toxins from the BP spill off the coast of Louisiana ended up in Minnesota pelican eggs.

In the twenty years since Hurricane Katrina battered the southern end of this river, more of it has been swallowed by a combination of rising sea levels and coastal subsistence, which jeopardizes many places where people live. At the northern end of the river, it has also become harder to ignore the signs that the Earth is changing. On my first full day in northwest Minnesota, a few of us waded through the Red Lake River, looking for water deep enough to swim. We noticed the bottom was lined with the shells of dead clams, which Gina said were killed by the heat. Several days later, the air above us turned hazy and choking from wildfires in Oregon, Montana, and Canada—a reminder that climate chaos knows no borders.

ONE PART OF A WHOLE

I was taught in college economics classes that people are inherently competitive and only act out of their own self-interest. That never rang true to me. I know too many examples of generosity among my friends, neighbors, and in my activist and faith communities. Living in a southern African village as a Peace Corps volunteer in my twenties taught me the communal concept of *ubuntu*, which means "I am because we are." My Irish ancestors had a similar word, *meitheal*, which conveyed community spirit and interdependence. Through two stays at Red Lake Treaty Camp, totaling about a month, I experienced the philosophy of interdependence at work. I saw people putting the community ahead of their own narrow interests, whether it was volunteering to clean the pit latrines or risking arrest to protect the water. People of all ages and genders cooked for each other, cleaned up after each other, and took turns staying up all night to guard the camp.

The community included people whose ancestors had come from Africa, Asia, Europe, and both Americas, but it was the camp's Indigenous women who set the tone. "We need people to come in a good way," said Sasha Beaulieu, leader of the camp and a member of the Red Lake Nation. That "good way" meant following the seven teachings of the Anishinaabe (the broader cultural group of which the Ojibwe are a part): respect, truth, honesty, wisdom, humility, love, and courage. When people fell short of these ideals or when tensions arose—which they do with any group of human beings—we were asked to pray about them in the sweat lodge. On a few occasions, when prayer and conversation were insufficient, someone was asked to leave camp because of disruptive or aggressive behavior. They were usually offered gas money and prayer. This was especially painful when the person was Indigenous. What struck me most about these incidents was that the safety and well-being of the group was prioritized over the feelings of an individual, which contrasted with mainstream American culture.

Although we were all addressed as "relatives," that concept was not used to deny differences in culture or experience, or to imply that we were all the same. I was struck how leaders like Sasha expressed both a fierce commitment to justice for their own people and an openness to their non-Indigenous allies, who were majority white in most camps. We came knowing very little of the history of the people and land we hoped to help protect. We had so much to learn, such as how to enter the sweat lodge and other ways of showing respect. We especially had to learn humility. As Tara Houska, founder of a resistance hub called the Giniw Collective, put it in a video for those considering coming to the frontlines of Line 3, "We need you to remember your gifts are just one part of a whole."

As someone accustomed to being a leader, I had to let go of my desire to have input or insider information about action plans that were intentionally kept secret until the last minute. At the large gathering at Shell River, I was humbled when Gina gently pointed out that I had cut in line for the booth selling T-shirts and local wild rice. I had approached from the wrong side and hadn't even noticed the people there before me. It was a mortifying mistake given the history of white people on Turtle Island, an Indigenous term for North America.

PROMOTING INDIVIDUALISM was central to the colonization of Turtle Island. As one commissioner of Indian Affairs reported to Congress in 1851, "It was our constant aim to do what we could to break up the community system among the Indians, and cause them to recognize the individuality of property." Privatizing land enabled companies and settlers to gain control of Indigenous resources, such as oil, minerals, and timber. In 1887, the Dawes Act was passed to divide remaining communal lands into private allotments. After visiting an Indigenous community, the law's architect complained, "There is no selfishness, which is at the bottom of civilization."[136]

The Red Lake Nation is one of only two reservations in the United States where the community owns 100 percent of the land for the benefit of all members. Holding fast to communal land ownership was an act of resistance in the face of ongoing pressure from settlers and the US government. Red Lakers also maintained their language and burial rituals in greater numbers than many other Indigenous nations.[137] Across most of Turtle Island, the allotment system fractured communities and enabled settlers to acquire land and do with it whatever was in their narrow, short-term interest.

Similar dynamics played out in many parts of the world as Indigenous people fought to protect the land that sustained them. Where colonists gained power, the result was often deforestation, monoculture agriculture, fossil fuel extraction, pollution, climate change, and the extinction of many species, whose habitats have been disrupted for profit.

During our interview on the bank of the Red Lake River, Gina remarked, "It's that individualistic thinking that's killing us. Before colonizers came, you could drink out of every river and every stream here on Turtle Island." Colonizers took Indigenous land and introduced a destructive and deadly way of doing things, one where a white man's right to be selfish trumped his responsibility to other people, let alone to the water or the fish. This was why Line 3 was being built through a huge concentration of fresh water, just so a tiny minority could have more than they needed, Gina said. "That's not how it's supposed to be." Noting that colonization also disrupted the gender balance of many Indigenous cultures, she added, "You might as well follow Indigenous women. Because you haven't for five hundred years. You might as well take a chance on us now and follow us instead of that individual path."

THE WORK OF THESE TIMES

Gina told me that when Indigenous people sometimes complained about being minorities in their own camps, she reminded them of the Seventh Fire Prophecy, which predicted that people of different colors would come together at this time of the crossroads. Rereading Robin Wall Kimmerer's version of the prophecy while at camp, I noticed that she predicted the journey would be challenging, which I hadn't really absorbed the first time I read it.

The first six Anishinaabe prophecies were also eerily prescient. Before European arrival, the first told the people that trouble was coming and they should move from their territory along the Atlantic coast, west to where food grows on water. The Anishinaabe heeded the warning and eventually settled amid the wild rice of the Great Lakes, spreading out in three related groups, the Ojibwe, the Odawa, and the Potawatomi. Then, two prophets foresaw the arrival of light-skinned people from the east. There was great potential if the newcomers came in brotherhood, but if they came in greed, the fish and waters would become poisoned. A subsequent prophet predicted that the people would suffer if they abandoned their own spiritual teachings and followed the religion of the newcomers.[138]

It is sobering to remember that the poisoning of the waters and the forced conversion of Indigenous peoples were not inevitable. They were choices, which is what prophets show us more than outcomes. Across what became known as the United States, hundreds of thousands of Indigenous children were ripped from their families and confined in violent institutions, misleadingly called "boarding schools." These institutions were pillars of the colonial power structure, which separated people from their land and the spiritual ceremonies that honored their connection to the land. The Seventh Fire Prophecy, the last in the series, predicts a time of Indigenous spiritual renewal. I felt honored to witness this at Red Lake Treaty Camp, as people from different nations shared the teachings of their elders and passed around books by Indigenous authors.

INDIGENOUS PEOPLE are not alone in understanding this as a pivotal era. Buddhist systems theorist Joanna Macy calls it "the Great Turning," a societal transformation on the magnitude of the Industrial Revolution. Instead of the scorched path, she calls our current system the "industrial growth society," which measures its performance "by how fast materials can be extracted from Earth and turned into consumer products, weapons, and waste." Instead of the "green path," she calls the alternative "a life-sustaining civilization" based on an understanding of the interconnection of all life. People are aiding this transformation in three major ways: slowing the damage to the Earth and its beings; creating new, less destructive systems; and facilitating "a shift in consciousness." Allowing ourselves to feel our grief for the clams baking in the river and the cities threatened by rising seas can aid this consciousness shift, along with spiritual and scientific insights about life's interconnection.[139]

Macy's three aspects of the Great Turning were all present during the Stop Line 3 campaign, even though it was most explicitly about slowing the damage to the Earth and its beings. "Every day we slow them down, the mussels get another day," said Winona LaDuke, one of the leaders of the campaign. She had invited three hundred people to gather under tall pines at a campground by the Shell River, which was slated to be crossed five times by Line 3. LaDuke got a chuckle from the crowd when she said that her Anishinaabe people had lived on this land for a long time with no problems. The rivers and fish were bountiful, and the maple trees provided sugar year after year. "You can still harvest the same rice from the same lake for ten thousand years. That's what you call a sustainable economy!"

The founder of Honor the Earth, LaDuke has spent decades dedicated to the second spoke of the Great Turning, creating less destructive systems. She farms hemp as an alternative to synthetic materials and supports other Indigenous hemp initiatives. LaDuke writes that until the 1920s, 80 percent of the clothing made in the United States was from hemp, with eleven hemp mills in Minnesota alone. Many chemicals used in clothing today are by-products of oil and gas, peddled to other industries to make petrochemicals more profitable. In contrast, hemp plants actually absorb carbon dioxide and make the soil healthier.[140] Remembering and adapting older ways of doing things can help us envision a less destructive future, another reason Indigenous perspectives are especially important today.

I THOUGHT I came to Line 3 to slow the harm to the Earth, maybe by sitting in the way of construction crews, the type of bold nonviolent direct action that some other camps were doing. Instead, much of my experience fell into Macy's third category, shifting consciousness. When I arrived at the Red Lake Treaty Camp, I was told that it was a prayer camp, and that practicing ceremonies on this land was itself a way of defending the treaty rights of the Red Lake Nation. The Ojibwe who fished in the rivers or hunted deer reminded us all that this land sustained people long before the Walmart was built five miles from camp.

We were invited to offer pinches of tobacco to the different rivers in the area to express respect and gratitude, even as we tracked the pipeline's progress and the depletion of local waterways by industry pumps. The fire that burned continuously at camp contained ashes saved from the sacred fire at Standing Rock, another reminder that we were part of a bigger ongoing story. Each of these practices expanded how I understood my connection to the Earth, making me feel I had been sent to the camp where I was meant to be, even when the lessons were difficult.

THE ROLE OF LAW ENFORCEMENT

Many water protectors at the Stop Line 3 campaign camps carried deep scars from the long history of state violence against Indigenous people, from massacres and forced removals to the everyday police harassment that Indigenous people have endured. Gina's father was one of the hundreds of thousands of Indigenous children torn away from their families, often by police, and sent to institutions designed to kill their culture.

Gina's father is also second cousin to Leonard Peltier, a leader of the American Indian Movement (AIM), which was founded in the late 1960s to connect and support Indigenous struggles. The Federal Bureau of Investigation (FBI) infiltrated AIM and, after a shootout at the Pine Ridge Reservation in South Dakota, accused Peltier of killing two of their agents. Despite evidence that the FBI coerced witnesses and falsified a ballistics report, Leonard Peltier was imprisoned from 1977 until Joe Biden commuted his sentence to house arrest, after decades of advocacy by Indigenous people and allies.[141]

"I know firsthand the dangers of what could happen if you speak out," Gina told me. As a child in North Dakota, she was bullied for being a Peltier and an "Indian." As an adult, she was abused by more than one male partner. Because of those experiences, she used to think that if she yelled at people and was volatile, they wouldn't mess with her. In recent years she learned that she is more likely to be listened to if she keeps her calm. "I have my moments, like, when an Enbridge commercial comes on the radio, I get my fucks out in the privacy of my home," she laughed. Although she gets angry, she now tries to make sure that anger doesn't consume her or escalate a conflict that could hurt the group. During a Minnesota rally against police violence after the killing of George Floyd, a police car knocked her down when she was serving as a peacekeeper. The crowd was angry, but Gina stood up and called for nonviolence, stretching out her arms to protect the police.

When Gina shouted over the fence to those who were on the Enbridge side of the line, she tried to challenge them on a human level. "I'm out here risking my life for your future, for your land, for your water," she told law enforcement, asking them to question why they were defending a corporation instead of people. To me, she said, "We're fighting for the pipeliners' lives, and they don't care about us. The law enforcement, they don't care about us. But here we are."

IN EQAT, there was sometimes tension between those members who wanted to acknowledge "that of God" in the police and those who saw the police as upholders of a violent and discriminatory system. The first group (often older white Quakers) occasionally thanked police officers for stopping traffic for a march, or for removing our handcuffs at the end of a civil disobedience action. It was often our younger members who bristled at these attempts at connection. They pointed out that, after the death of George Floyd, Philadelphia cops had led marchers to an enclosed area and then tear-gassed people—including some of our own members. Some argued that people of color or others who felt vulnerable to police violence might not feel safe at our actions if we seemed too friendly with law enforcement. In contrast, I once attended an Indigenous-led action in DC where a few white people screamed insults at police, concerning people of color who feared that such performative anger could provoke law enforcement.

These tensions also existed at Stop Line 3 camps. Not all Indigenous people spoke as gently as Gina, though more than in most groups I had been part of, the philosophy of interrelatedness at Red Lake Treaty Camp allowed space to appeal to the humanity of individual law enforcement or Enbridge workers, while also being clear about the brutal system they served.

THE ROLE of law enforcement in upholding the industrial growth society was blatant along Line 3, where Enbridge gave millions of dollars to local sheriff departments to defend the Canadian company's interests.[142] From the frigid months of Minnesota winter through the summer of 2021, more than nine hundred water protectors were arrested for nonviolently resisting the pipeline's installation. Some were simple sit-ins, as when Winona LaDuke and six other elder women sat in lawn chairs on an easement, preventing Enbridge from accessing the Shell River—at least until the women were arrested. In other nonviolent actions, people locked themselves to excavators or drills, which delayed construction for longer. To discourage this sort of thing, charges escalated to felonies, with prosecutors threatening long jail terms. Law enforcement violence, which had been present throughout the campaign, also escalated over the summer. Far from media hubs, violent incidents received little national press coverage, minimizing the opportunity for water protectors to build public sympathy through their sacrifice.

When the drill arrived at the Red Lake River, water protectors came from other camps, including Tara Houska, whose women- and two-spirit-led Giniw Collective had led many arrestable actions along Line 3. Crossing the line that divided the green grass of the Red Lake Treaty Camp from the machine-filled Enbridge worksite, Tara was peppered with rubber bullets at close range. Other water protectors were also shot with rubber bullets and pepper spray when they approached the machinery. I went home just before these events, but I watched Facebook Live in horror as uniformed officers slammed one young person's head into a large machine. His limp body fell out of view. I saw a photo of an Indigenous woman slammed to the ground while in spiritual ceremony near the Enbridge driveway. Many water protectors were arrested.

In vivid contrast, Minnesota law enforcement did not arrest a single Enbridge employee for three aquifer breaches, or for spilling ten thousand gallons of drilling fluid over twelve river crossings. The state fined Enbridge only a tiny percent of the $4 billion project.[143]

BACK HOME in Philadelphia, I felt grief-stricken about what was happening at camp. Knowing that action is medicine for despair, I fundraised for bail and legal expenses for friends there, and organized a solidarity civil disobedience action in Philadelphia. Twenty-eight people showed up at short notice. Some had been to Line 3, or their adult children had. Many were from EQAT. Six people risked arrest with me, which in this case meant refusing to leave the vestibule of a downtown TD Bank, one of the financers of Line 3. The bank chose to shut down its own center city branch rather than instruct the police to arrest us. They knew that in a visible urban area, arrests might increase attention to what we were doing, which the bank did not want.

REGROUPING

I believe that any nonviolent direct action has an impact, even when it is mostly felt by the participants. As my ad hoc crew held space inside the TD Bank entrance and out on the sunny sidewalk, I was moved to hear why Philadelphians had shown up in solidarity. Invited to give short speeches, they took turns, speaking of Ojibwe rights, the waters of Minnesota, their own children's future, and the money that many of us invest that ends up financing projects like Line 3 without our knowledge. One said, "Nature is making the connections. We need to catch up." On a livestream of the action, someone said, "We are all connected," observing that Philadelphia was currently smoky from western wildfires. I was heartened that the Stop Line 3 campaign was helping people to feel these connections, shifting consciousness from a distance, even though we hadn't stopped the pipeline, which was almost complete.

In August, I drove my Prius Prime back to the Red Lake Treaty Camp. Only a few weeks after the pipeline was installed there, I could feel the weight of the trauma people had experienced.

There were now camp members guarding a new wire gate at the driveway entrance. I proved I belonged by calling one of the camp dogs by name, the only mammal whom I recognized for the first few days, until Gina and other friends returned. I pitched my tent near where I had before, but moved it downriver a few days later, since Enbridge was now running bright lights and noisy machines all night long in a rush to complete the pipeline by summer's end. I soon felt drained from a lack of sleep and developed a terrible cold. I tried to make myself useful by helping to collect and sort the possessions left behind by water protectors who had come for the pipeline's installation and then never returned after the violence and arrests.

As the whole camp struggled to regroup, a new wave of Indigenous youth organized a prayer action at the Red Lake River, over the installed pipeline. They hoped the police wouldn't interfere, now that the damage was done. Most people sang prayers on floating wooden rafts that had to be pushed upstream to the site. I was one of six white allies or "accomplices" asked to stand as a safety barrier between the police and several Indigenous youths who climbed the Enbridge easement to sing and drum through the chain-link fence between the noisy worksite and the river. In the long skirt I had been taught to wear for ceremonies, I prayed fiercely, holding one of six thin wooden shields that were covered with reflective metal, so that the police would see themselves if they looked at us. I was moved by the drumming and plaintive singing around me and later asked what the words meant. A young Indigenous man with long thin braids told me that when he was asked to sing a song, he consulted the Creator for guidance, and the words that came to him were a prayer for those doing the harm. "They don't know any better," he said quietly.

The message of interconnection was reinforced by the presence of our animal relatives at the action. An eagle flew high over the river before we began. An otter bobbed surprisingly close to the first raft as a turtle watched from a protruding rock. "Of course, those are our clan animals," one Red Lake elder explained. I was not raised with a worldview that allowed for animals attending prayer protests, but as I waded between the otter and the turtle on my way to the Enbridge fence, I remembered Robin Wall Kimmerer saying that those of us moving toward the green path are aided by other living beings who want to live, too.[144] I was told that a hummingbird—a species so fast it can carry messages from the spirit world—had flown into the tent where the Indigenous youth were planning the action. To remember this lesson, I got my first tattoo at age fifty-nine—a hummingbird on my shoulder, a reminder of Spirit that felt less culturally specific than the Indigenous symbols some water protectors were having tattooed.

WHEN I left Red Lake Treaty Camp for the second time, it was with three other water protectors in my Prius Prime. We shared music from our different cultures as we drove to a large march in St. Paul, pausing to visit the narrow section of the Mississippi where Gina had been arrested a few weeks earlier. Looking down at the winding river surrounded by knee-high greenery, I saw a muskrat dunk under water and reappear three times. It felt like another teaching. The humble muskrat plays a crucial role in the Anishinaabe creation story by sacrificing himself to find soil at the bottom of the sea for the first helpless human to stand on after she falls from the sky. The turtle volunteers his back to hold the soil, which grows into the continent of Turtle Island. Skywoman encourages the flourishing with a dance of gratitude and the seeds she brought to share with all. Each species in this story offers what it has.[145]

Each of us has a gift that is needed now in this time of the Great Turning. When the scale of our problems feels overwhelming, I remember that, like the muskrat and the turtle in the Anishinaabe creation story, I just need to play my part, feeling gratitude for the many others who are playing theirs. My experience at Line 3 strengthened my appreciation for those whose gifts were around prayer and healing, which are part of what Joanna Macy calls "facilitating a shift in consciousness."

HEALING

One of the people I stayed in touch with after leaving camp was Gigi Nathan, who had come to Red Lake Treaty Camp to pray after watching video of the brutal installation of the pipeline. A Tsimshian and Haida healer, whose people are from Ketchikan, Alaska, Gigi's healing powers helped me recover when I got sick after Enbridge's scorching lights kept me from sleeping. A few months later, Gigi started a weekly healing group to address the trauma many water protectors had experienced. It was still meeting when I visited her in Seattle almost a year and a half after we left Line 3. She told me how her healing gift developed when she reconnected with Indigenous ceremonies during her mother's cancer.

Gigi's mother was a boarding school survivor, who "partied all night" when she was young. Gigi witnessed her mother's transformation as she overcame addiction and became a therapist. Although they didn't have much money, she raised her children in the suburbs to shield them from the violence and addiction present on the reservation. "I was my mother's most colonized child," said Gigi, who held corporate jobs when she was young. When Gigi started caring for her dying mother, she realized "there was a different way to be" than how she had been living in main-stream culture. Gigi started seeking out Indigenous ceremonies, like the sweat lodge, where she saw and felt the presence of her female ancestors smiling at her with support.

Twenty-two months after her mother's death, Gigi's son was hiking on Deer Mountain in Ketchikan when he disappeared. It was November 2015, and the Alaska weather turned quickly. The state police spent four days searching for him, then called off the search due to snow. Gigi and her family continued to look for him for a month. She was shattered and exhausted, but remembered how her mother's death had spurred her healing journey. "The only thing that I knew was that it was possible to heal," she told me. "The only thing that made sense was the Red Road. Different Indigenous ceremonies. Outside of that, nothing made sense."

During the winter months of not knowing what happened to her son, Gigi prayed. She also took an REI course on rock repelling so she could keep searching in the spring. When she headed into the mountains again, her sister called the state police and told them that they could not let a mother be the one to find her son. They resumed the search and found some of his ribs, confirmed by a DNA match. He had fallen off a cliff. Gigi had to raise money for cremation and didn't get his ashes until more than a year after his disappearance. After this shattering experience, she became a Sundancer, a participant in the special Indigenous healing cer-emony that was also important to her mother. After dancing to exhaustion in the hot sun, Gigi saw her son's smiling face and felt reassured that she was where she was meant to be.

Through her own slow recovery, Gigi started supporting the families of Missing and Murdered Indigenous Women (MMIW). Indigenous people, especially women and girls, are disproportion-ately targeted for violence, including rape and murder. Federal

Indian law makes it hard to prosecute non-Indigenous people who commit such crimes on reservations, which contributes to the fact that many relatives are never found. These problems often get worse around the temporary worker camps that spring up near construction projects like Line 3.[146] Supporting MMIW families and accompanying them on prayer marches gave Gigi motivation to continue her own healing, so she could help others. "No one else understood what they were going through. How could you?" Gigi recalled.

LISTENING TO Gigi, I realized how cruel it was that the US Congress banned the Sundance and other sacred ceremonies in 1883, just when Indigenous people were experiencing so much trauma.[147] I acknowledged the difficulty of feeling connection with people who have done so much harm, and told her that something she had shared in Minnesota had stayed with me. "You can't just pray for someone who has been hurt," Gigi had said. "You also have to pray for the person who hurt them, and the people they will hurt out of their own pain."

Gigi shared what helped her learn this lesson. There was a ceremony honoring MMIW families before a women's march in Seattle, but the ceremony was delayed due to audio problems. When they were fixed, the organizers tried to jump ahead in the program, but an ally insisted the MMIW families be allowed to finish what they came for. When the march finally started, one hundred thousand people rushed into the road where the Indigenous families were supposed to walk in front. Gigi was surrounded and separated from the group. Wearing traditional attire, she asked people to let her pass, and they made rude comments like, "Well, you all should have started the parade on time." One woman helped her to rejoin her group, but when they turned a corner, another rowdy crowd streamed into the street, oblivious to who they were displacing. Gigi was so angry, she stopped praying, which was why she had come.

"It probably took me twenty minutes to get my shit together before I could pray again," she recalled. I told her twenty minutes wasn't bad compared to most people, and we both laughed. She said that incident was important in spurring her path. Afterward, she made four hundred prayer ties for people who went missing under the age of twenty-five, and also said prayers for the murdered.

"My ancestors were telling me that I also have to help white people heal," she recalled. "Why would I help them heal? They're the ones killing us," she responded. "How are you going to get them to stop killing you if they're not healed?" the ancestors asked. "Okay. Valid point," Gigi recalled with a laugh. She said that healing was her hope and prayer for white people, which she saw as connected to the healing of the Earth. I thought of my Irish ancestors and wondered if being able to heal their trauma from the famine of the 1840s might have helped more Irish Americans resist colonization and racism in the United States.

I asked Gigi to share more about her healing practice and whether it was anything like Reiki. She said no, that when she heals people, it is through medicine that comes from within her. What she puts into her body creates it. "Transforming my trauma, and healing my trauma creates it. So, the more that I do, the more medicine is created that I have available for other people." Gigi said that "disconnection" was often a core issue for people, whether their symptoms were mental or physical. "Today, the person that I helped was so disconnected that they were in a lot of pain all over their body." She mentioned that it was a non-Indigenous relative who was also fighting the colonial system. Using her hands, Gigi "took out a lot of what was weighing them down, but also reconnected things for them." The relative was calm and out of pain afterward, she said. I noted that in my years of receiving health care from the University of Penn medical system, no one ever referred to me as a relative. Gigi nodded. Between us sat a bowl of smoldering sage and sweetgrass, medicines from the Earth.

LESSONS FROM LINE 3 AND BEYOND

A Quaker friend went to the Red Lake Treaty Camp after the Enbridge machines were removed and the scorched earth next door was abandoned. She told me that on a solitary walk past the former worksite, she could literally hear the cry of the Earth where it had been violated. It was a powerful reminder that just because a scar is out of sight doesn't mean it is healed.

In addition to emotional scars, many water protectors were dealing with legal charges long after the pipeline started pumping oil. Eventually, a Minnesota senior district judge dropped lingering charges against Winona LaDuke and two other Ojibwe water protectors. The judge acknowledged that she had grown up with negative stereotypes of Native Americans and only gradually learned the history of the injustice they faced. Acknowledging the water protectors' treaty rights and their spiritual beliefs about protecting the Earth, she said, "To criminalize their behavior would be the crime."[148] Apparently, part of what impacted the judge was having so many Indigenous defendants come before her bench, perhaps wearing down her illusion of separation from them.

Winona told *Democracy Now!* that she cried when she heard the judge's words. The story reminded me that individuals who serve the pillars of power can be moved to change.[149]

One of the most moving speeches I had heard at the gathering at Shell River was by an Ojibwe man who was working for Enbridge when an Ojibwe woman he recognized led a nonviolent action at his pipeline worksite. He quit his job and joined them the next day. At Shell River, his singing and drumming played a powerful role in connecting the group while expressing grief over what was happening to the land. The campaign needed his gift.

AFTER MY time at Line 3, I heard EQAT's discussions about policing in a new way and started adapting a framework developed by Niyonu Spann, founder of Beyond Diversity 101 and director of Co-Creating Effective and Inclusive Organizations. In her transformational workshops, Niyonu talks about three circles or layers to identity.

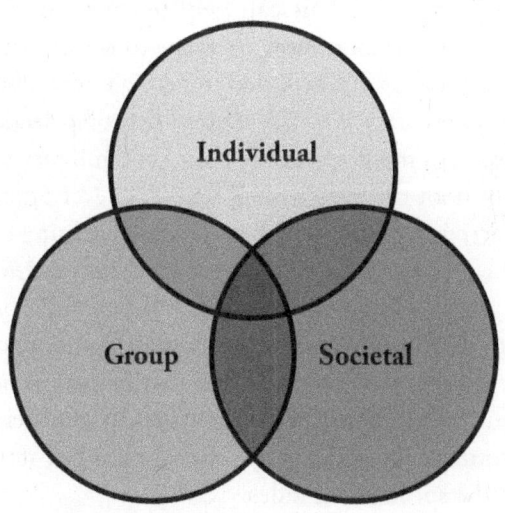

There is the individual level, the group level, and the societal level. All exist simultaneously. Because our society encourages us to see ourselves as individuals, those with privilege often have a harder time recognizing group and societal dynamics than those who are targeted by those systems. Some people's survival depends on understanding the role of police, for example, while others don't think much about law enforcement.

When I use the three circles to facilitate this discussion, I put people in small groups, often by generation, and ask them to reflect on how they were taught to think about the police. After they report back, I point out the ways class, race, region, and generation can influence both our individual and group experiences. If the group hasn't already acknowledged the societal role law enforcement plays in defending private property and the status quo, I explain that. I emphasize that all three levels exist at the same time. Recognizing this can help us make more conscious and strategic choices about how we relate to law enforcement.

A great example of this occurred in Serbia under a brutal dictatorship. Pro-democracy activists realized that they needed to build relationships with members of the police and military to convince them not to shoot the protestors if they were ever ordered to. For two years, activists went to the bars where they hung out and listened to their complaints. They sent cigarettes and food to the underpaid soldiers. The strategy worked. When millions of people flooded into the capitol in protest, the military refused orders to shoot their neighbors, a key moment that helped to bring down the dictator. Civilian activists had removed an important pillar of power by strategically shifting people who did not start out next to them on the spectrum of allies.[150]

Years ago, I attended a Philadelphia protest against fossil fuel train shipments through our city. It was a small crowd, and some participants noticed a man on the periphery listening to the speeches. They decided he "looked like a cop," which made them suspicious, so I walked over to speak with him. He said that he was a cop, who was worried about the safety of such trains, especially because his daughter lived in a neighborhood where they passed. I've heard that some members of law enforcement acknowledge that if a pipeline breaks, they will be the first people exposed to dangerous chemicals. As far as I know, few pipeline fights successfully turn cops into allies. Of course, such organizing is extra hard if your community has experienced police violence or if a corporation is giving the local police department millions of dollars to defend corporate interests.

I thought more about these dynamics in the lead-up to the 2024 election, when I worked with Daniel Hunter, who created training materials in case Trump won a second term. Lessons from places like Serbia suddenly felt much more relevant to the United States. After the election, it became clear that judges, prosecutors, and others working within government would have significant choices ahead of them. Being able to speak to the humanity in those people became not just a spiritual ideal important to Quakers but an important political strategy. While the left has often been quick to denounce people who don't share all our ideas or practices, it feels like a good time to remember than many people working within the pillars of power could become allies, under the right circumstances. In fact, we need them to if we hope to steer away from the scorched path. Seeing them as relatives could be a start.

CHAPTER SEVEN

JUST TRANSITION

For nearly sixty years, coal mining was a major industry on the Navajo Nation, which spans Arizona, New Mexico, and Utah. The story of coal's decline there shows what can happen when different groups of people weaken the pillars of power supporting an industry that is terrible for the climate. At the same time, coal's demise illustrates why the Great Turning isn't just about stopping the harm of the industrial growth society but also about transitioning to a life-sustaining civilization for everyone.

Home to burial grounds and other sacred sites, the plateau called Black Mesa has spiritual significance for both Hopi and Diné (who colonizers called Navajo). These neighboring nations have often been pitted against each other,[151] but in 2001, youth from both communities decided to work together to protect their shared aquifer, calling themselves Black Mesa Water Coalition. Peabody Coal was using millions of gallons per day, drying out springs that people relied on to graze their sheep. The water protectors educated their people at chapter houses, the local governing bodies, until the Navajo Nation government was convinced to put limits on Peabody's water use in 2003.[152] Pressure was also building from other directions. The Sierra Club won a lawsuit that, combined with the Navajo Nation's stand, made it harder for Peabody to foist the true cost of coal onto Indigenous people. The Mohave Generating Station and the Black Mesa mine that supplied it both closed at the end of 2005.[153]

It took fifteen more years to close another major coal opera-tion, the Navajo Generating Station, and the Kayenta mine that supplied it. Meanwhile, the Black Mesa Water Coalition started pushing for a transition to renewable energy, especially solar, using land that was already despoiled by coal. They pointed out that, unlike environmental advocates in some places, they had family members who were losing coal jobs. They wanted to help create alternatives that would keep their people on the land of their ancestors. After a green jobs march that involved many organiza-tions, the Navajo Nation passed green jobs legislation in 2009, the first such law by any tribal government.[154]

In 2011, Wahleah Johns, then the co-executive director of the Black Mesa Water Coalition, spoke to a group of California women working on sustainability. After describing the land of her home, she asked, "How are we connected? Does anyone know how Black Mesa is connected to Los Angeles?" The audience guessed water and power; Wahleah nodded. Southern California had long used energy produced from Black Mesa coal, transported with Navajo water, while 50 percent of Diné still didn't have electricity and 40 percent didn't have running water. "Even though we are located hundreds of miles away, there is this connection," she said. She added that they were also connected as mothers, grandmothers, and sisters. She encouraged the women to use their power for a just transition that included Indigenous people.[155]

THE CONCEPT of "just transition" grew out of the labor move-ment, as workers in the fossil fuel industry realized they could be hurt by the transition to clean energy if there weren't programs for retraining and compensating industry workers. Just transition has since grown to include other justice issues, like who has access to energy and who profits from energy production. For many, it is also important that the transition addresses the harms done by the old energy economy, from cleaning up industry pollution to helping people cope with climate chaos. Politically, connecting these issues can help to build broader support for change, as it did in New York State, where unions have been part of the coalition. Even before the term "Green New Deal" was popularized nationally, local communities were exploring the local job potential of solar and wind, as well as other financial benefits. These developments helped inspire EQAT's Power Local Green Jobs campaign in 2015.

The first solar field on the Navajo Nation was built in 2016, but coal still provided nearly 80 percent of the Nation's revenue that year.[156] Despite residents testifying that the coal operations correlated with increased asthma and cancer, the tribal government lobbied to keep big coal operations going, citing the need for jobs.[157] But forces outside the Nation were also at play. California and Nevada—long destinations for the energy produced with Navajo coal—passed laws prioritizing cleaner energy, which was gradually becoming cheaper.[158]

When the owners of the Navajo Generating Station themselves decided to close, water protector Nicole Horseherder livestreamed the demolition of the station's enormous smokestacks to ten thousand viewers, narrating in both English and Diné. Climate activists nationally celebrated the end of one of the largest carbon emitters in the United States, but the 2020 event evoked complex reactions among Diné, some of whom were given the option of moving away for work. Director of the Navajo Nation's Division of Natural Resources at the time, Bidtah Becker, later testified to a US government subcommittee: "One of the key lessons for the Navajo Nation is to not recreate the situation where the Nation is at the mercy of others for energy development on the Nation when the livelihoods of Navajo people are at stake . . . the Nation has learned the difficult way what ensues when the Nation does not have a management voice."[159]

Activists also wanted new energy industries to be different. As Nicole Horseherder told *High Country News*, "We want to make sure that there's clear guidelines as to how to operate projects on the reservation, especially for outside entities, but also for inside Navajo developers." She hoped the new solar industry would be not just cleaner but less colonialist in the way it operates.[160] For Indigenous people, this is another key principle of a just transition. Diné history shows how colonialism has served as another pillar of power supporting extractive and polluting industries.

WHAT'S COLONIALISM GOT TO DO WITH IT?

Colonialism involves one nation claiming control over the land, people, culture, and/or politics of another nation, often for the exploitation of resources. Like racism, it is justified by prejudiced ideas, which are used to normalize violence and exploitation.[161] In the colonization of Turtle Island, religious prejudice was used before the ideas of racial prejudice solidified.

The Roman Catholic Church developed the Doctrine of Discovery to support European monarchies seeking to colonize lands beyond Europe. The doctrine was formed through a series of papal pronouncements, the last of which was written soon after Christopher Columbus returned from his first trip to the Caribbean in 1493. Although he had been looking for India, Columbus wrote to the Vatican and his Spanish financiers, reporting that he had claimed for Spain many islands inhabited by unarmed people with many valuable resources.[162] Asserting that "the extension of the Christian realm" would please God, the papal bulls gave theological justification for Christians to enslave non-Christians and take their land through brute force.[163]

SPANISH CONQUEST spread, eventually reaching the vast lands of the Diné, which included arid desert, alpine forest, and high plateau. Although Europeans began settling in the area, the Diné defended their independence for nearly three centuries. Meanwhile, the US Supreme Court cited the Doctrine of Discovery in an 1823 decision, using the doctrine to justify the continued expropriation of Indigenous land. By the mid-1800s, the "Anglos" had replaced the Spanish in what we now call the Southwest. They justified the displacement of Indigenous people with the idea of "Manifest Destiny," the assertion that God intended for the United States to expand across the continent. When the US Army arrived in 1864, soldiers sealed Diné water sources, burned thousands of peach trees, and slaughtered sheep to dislodge the Diné from their land. Without food and water, well over eight thousand people were forced to march hundreds of miles under brutal conditions. Those who survived the Long Walk were held for years at an internment camp east of the Rio Grande.[164]

Some US officials hoped to relocate the Navajo to "Indian Territory," in what is now Oklahoma, but they were dissuaded by Barboncito, a Diné spiritual leader from Canyon de Chelly. Through translators, Barboncito spoke about the importance to the Diné of returning to Diné Bikéyah, the land of their grandparents. He explained that at home, the people could talk to the snakes, but in the strange land where they were being held, the snakes didn't know them.[165] He told the US officials that the Navajo would live in peace if they could go home. General William Sherman, famous for his scorched-earth approach to Confederates during the Civil War, relented, and a treaty was signed. Against all odds, several thousand Navajo made the long walk back to their territory, which under the treaty was smaller than it had been, but still larger than any other reservation in the United States. In the century and a half since, the Navajo Nation has grown to over 17.5 million acres.

I visited the Navajo Nation Museum and quietly read the mantra painted on white walls: *T'ahdii kqq" honiidlq'. T'ahdii kqq" honiidlq'. T'ahdii kqq" honiidlq'.* We are still here. We are still here. We are still here." The treaty was on special exhibit for its 150th anniversary, each page of slanted cursive matted under glass. I was moved to see the names of twenty-nine Diné leaders who marked an *X* for their signatures. I was also sobered to read the conditions of their return to Diné Bikéyah. One provision stipulated that the Navajo send their children to American boarding schools, which some today call "concentration camps." Another provision stated that the Navajo "will not in future oppose the construction of railroads, wagon roads, mail stations, or *other works of utility or necessity*" on their territory.[166]

In the 1920s, Standard Oil wanted to drill on Navajo land. Traditional Diné leaders resisted. US officials intervened and handpicked a council of boarding school educated Navajo who allowed Standard Oil to drill.[167] The United States went on to replace traditional leaders in many Indigenous nations with similar councils that were friendlier to corporate interests. Unsurprisingly, the wealth generated by oil drilling mostly flowed off the reservation.

THE UNITED STATES continued to violate Diné sovereignty. In the 1930s, the US Geological Survey worried that overgrazing on the Navajo Nation was causing soil erosion and silting up the Colorado River, which runs around the Nation's northwestern corner. They feared that silt would impede downriver Hoover

Dam, which was being built to provide electricity to California. With no concern for the Navajo, the Bureau of Indian Affairs (BIA) orchestrated the slaughter of hundreds of thousands of sheep, goats, and horses on the Navajo Nation.[168]

Diné accounts of these events feature the sound of women weeping for animals they considered relatives. People hid what animals they could. Medicine men led special prayers and ceremonies. Women gave outraged speeches at community meetings, confusing the white men who didn't understand that most sheep were owned by women, who were powerful in this matrilocal society.[169] Those Diné who tried to interfere with the slaughter were jailed. The people reviled the councilmen and translators who collaborated with the BIA, as well as its leader, a white social worker and artist who saw himself as their ally.

Traditionally, large Diné herd owners were expected to help needy kin, something they were no longer able to do.[170] Hunger forced the destitute to seek paid work in the white man's economy. This helped open the door to future extractive industries, as people found it harder to sustain themselves from the arid land without their animals. Diné said that killing the sheep had driven away the rain clouds, and the land was in mourning.[171]

Before the twentieth century, Diné culture considered coal mining taboo.[172] By the 1960s, the Navajo Nation Tribal Council was issuing leases for strip mining on Black Mesa. Peabody Coal got a flagrantly exploitative mineral lease by secretly putting the white negotiator for the Hopi on its own payroll.[173] The company also lobbied Congress to expel people from the land by exaggerating a land dispute between the Navajo and the neighboring Hopi, who had shown themselves more resistant to coal mining and the tribal council system. The divide and conquer strategy worked. To make way for coal mining, vegetation was bulldozed, livestock was confiscated, and at least ten thousand Navajo and one hundred Hopi were driven off Black Mesa, more people than were forced on the Long Walk.[174]

These stories illustrate how colonization contributed to building the fossil fuel economy, but they also explain why many Indigenous people worry about what will replace fossil fuels. Solar panels and electric vehicles need minerals, some of which are plentiful on Indigenous lands. In recent years, nuclear energy has drawn renewed interest from both the Biden and Trump administrations,[175] while Diné are still suffering from the legacy of the first uranium boom. Visiting Diné Bikéyah, I found uranium to be an especially sobering reminder that the crisis of the Earth is about more than coal, oil, and gas.

INVISIBLE DANGERS

"There used to be a nice little meadow there for the sheep to graze," Edith Hood recalled as we sat under a wooden pavilion her small community had built near the eastern border of the Navajo Nation. A Diné woman in her sixties, she pointed to a tan hill that was a sacred burial site in the time of her grandparents. Then, in the 1940s, the Cold War drive to amass nuclear weapons initiated four decades of uranium mining on Navajo land. The US government guaranteed a profitable price for uranium to private mining companies but did not require them to issue masks or other protective gear to their workers until long after the government and the companies knew the deadly risks. When demand for uranium dried up in the 1980s, the companies abandoned over five hundred mines without even bothering to put up fences or signs warning about the material's extreme danger.

"No one told us," Edith said of the danger. Diné grazed their livestock on uranium waste piles and used the yellow dirt as foundation for houses and corrals. Children played on the unmarked hills and sometimes in mine shafts. Some vents from the abandoned mines came up near people's houses. Under the wide cobalt blue sky, Edith pointed to where there had been filtering ponds with no fence to keep out their animals, some of which were found to have yellow insides when they were butchered. "Of course, it's in the water that gets into the plants and then gets into the animals and gets to the humans," she explained as I noticed the wind spreading the dusty top soil.

First Diné miners, then large numbers in the wider population, started suffering from cancer, diabetes, kidney disease, and chronic inflammation.[176] Edith, who was a cancer survivor, said the soil near her home was finally tested in 2007 and found toxic. When I met her, she had spent more than a decade organizing with her cousins and neighbors. She testified to Congress and tirelessly lobbied the EPA to completely remove the radioactive waste. Instead the EPA moved the worst soil just down the road. Another hill of waste, a short walk from where we sat chatting, was covered with dirt and planted with scruffs of green. An insubstantial wire fence bore a sign that read, "Danger: Abandoned Uranium Mine. Keep Out. *Ba' ha' dzid—Doo Ko' ne' na' adaa' da.*" The fence was the result of the community's advocacy. Many uranium sites aren't marked at all.

As we spoke, a small plane repeatedly zigzagged over the sky, the letters *EPA* printed on its underbelly. I later learned it was measuring gamma radiation, which was still "elevated." When two of Edith's cousins joined our conversation, the women complained that both the US EPA and the Navajo Nation EPA were not responsive enough to their concerns. When I asked why,

Edith said, "Because it's not in their backyard. It's in your yard, you do something."

A few years after my visit, the EPA finally planned to remove the uranium-contaminated soil in Edith's community, but the agency wanted Edith to move during the seven-year project. Determined to stay on their ancestral lands, Edith counter-proposed that they move to a nearby mesa. The EPA rejected this option when the Navajo Nation said it would not provide electricity there. Undeterred, Edith reached out to architects at University of New Mexico who specialized in Indigenous design, and together they planned a solar-powered cluster of modestly sized homes with state-of-the-art water reuse and doors facing the rising sun, a hallmark of traditional Navajo homes. "I feel empowered with those people," Edith told a reporter from *The Washington Post*.[177] It struck me that empowerment was a sign of true allyship.

FINDING ALLIES helped with another problem Edith's community faced, the lack of research and good data on the health impacts of uranium exposure. When they heard about environmental justice grants for such research, they reached out to Johnnye Lewis, a toxicologist at the University of New Mexico, who had a track record of getting grants. Johnnye agreed to collaborate, while helping the community grow their capacity to get grants themselves.

I visited Johnnye in Albuquerque in 2018 to hear about the Navajo Birth Cohort Study, which began because people were concerned about how uranium exposure was affecting their children. Although a long-term project, she told me that preliminary data showed that over 30 percent of Navajo babies have uranium in their urine by their first birthday, some on their first day of life.[178] For a researcher, measuring the impacts of this exposure was complicated by widespread poverty, which carries its own health risks. She also noted that it was hard to prove exactly how the contamination occurred, since uranium is in the soil, but when kicked up, travels in both dust and water, along with arsenic and other heavy metals that can alter DNA.

I explained the idea of the illusion of separation and asked if that applied here. Johnnye nodded. "Forty percent of the surface water in the Western US has been contaminated with mine waste," she said. That means the issue affects everyone. Indigenous communities are disproportionately in the West, and their economies are disproportionately reliant on natural resources, often without being part of the decision-making. Johnnye recalled thirty years earlier when she started working on the remediation of uranium mills. "What was stunning to me was how many of them were on native land. Then you start meeting the people." She pointed to a street sign near the outdoor table where we were sitting to illustrate just how close people's homes were to the uranium. "You don't find that in other communities."

Johnnye named "lack of connection" as one reason many people don't know about these problems. She added that contamination is also spread through windstorms, which are intensifying due to climate change. "You have to wonder what is in that wind," she said. I remembered how Hurricane Sandy kicked up old pollution in New York, though through water instead of wind.

THE HALF-LIFE OF MISTRUST

The largest spill of radiation-contaminated waste in US history occurred in 1979 in Church Rock, just outside the border of the Navajo Nation, a few miles from Edith's community. Weeks before new environmental regulations would take effect, the United Nuclear Corporation had installed a tailings pond on soil deemed unstable by state inspectors and then filled it higher than it was supposed to, merely patching cracks that appeared in the walls.[179] When the pond burst, over 93 million gallons of radioactive waste flowed into the Puerco River, killing sheep and crops along the banks.[180] Both Edith and Johnnye pointed out that the Church Rock accident didn't get as much press as the partial meltdown of the Three Mile Island nuclear reactor a few months earlier.

Near Harrisburg, Pennsylvania, Three Mile Island had 2 million people living nearby, many white. That accident prompted an evacuation order, a $1 billion cleanup, and follow-up health studies conducted by the EPA, the Nuclear Regulatory Commission, and several other agencies.[181] The Church Rock spill received much less attention. Federal, state, and Navajo Nation bureaucrats argued with each other and the company (now owned by General Electric) over who should pay for cleanup, which never fully happened. Edith's community commemorates the spill every year with a walk to the site, where they say healing prayers. They feel everyone else has forgotten.

Decades after the Church Rock spill, in 2015, a Diné doctoral student named Tommy Rock conducted a study of drinking water in Sanders, Arizona, along with Chris Shuey, director of the Uranium Impact Assessment Program of the Southwest Research and Information Center. Their tests confirmed what the Diné community along the Puerco River already suspected: The radioactive waste from Church Rock had slowly migrated across the state border and had contaminated their water supply, including the local school's. After retesting to confirm their findings, the researchers discovered that the Arizona Department of Environmental Quality (ADEQ) had data dating back to 2003 showing uranium in the Sanders water significantly above the allowed limit. The agency had done nothing to publicize or remediate the contamination for over twelve years.[182]

When the new research finally pushed them to act, the ADEQ sent a letter to residents saying the drinking water contained high levels of uranium but "does not pose an immediate risk to your health."[183] Their rationale was that the risks were long term, not immediate, even though their inaction had subjected people to long-term exposure. "I was livid," Chris told me, especially when he learned that the EPA had approved ADEQ's misleading letter. A combination of media attention, scientific data, and fierce advocacy got clean water for the people of Sanders, but years later, Chris was still disgusted by the indifference of the regulators.

Hearing the illusion of separation in the story, I asked Chris what he thought was behind the apathy. Chris, who is white, said regulators might have asked questions earlier if it had been their own families drinking the water. Their physical distance was combined with "this overwhelming notion that somehow Indigenous people are expendable. Or even more racist is this notion that they are poisoning themselves." Although he believes rural people in general are discounted, he's heard white people blame Indigenous health problems on people drinking too much alcohol or eating too much fry bread, stereotypes that make it easier for them to discount legitimate grievances. A dismissal he's often heard is, "Well, if you just got a job, you wouldn't have these problems." Chris shook his head in frustration.

GIVEN THE history of white indifference to Diné lives, it's understandable that many are anxious about renewed interest in uranium, spurred by talk of nuclear energy as a climate-friendly alternative to fossil fuels. There is hot debate about this. Once they are up and running, nuclear reactors contribute less to climate change than fossil fuel plants and cause less pollution—when there are no accidents. Nuclear accidents, when they happen, are especially hazardous. Of concern to many Diné, new nuclear plants would require more uranium. I asked Johnnye Lewis if there was a safe way to mine uranium, and she said, "Not that I've seen." Even the latest technology, which extracts uranium without pulling up so much ore, impacts water quality in ways that mining companies haven't been able to undo.

After decades of health impacts from uranium, the Navajo Nation outlawed uranium mining, processing, and transportation across its territory. But this is one of those areas where Indigenous sovereignty is still limited by the US government, which regulates national highways. Two such highways run through Navajo territory. In August 2024, dozens of Diné—including Navajo Nation president Buu Nygren and his wife—held a rally and blocked traffic after a company began driving trucks of uranium ore from the edge of the Grand Canyon to a uranium mill right outside the Navajo Nation, in White Mesa, Utah. Most of the three hundred mile route crossed the Navajo Nation, but the company said it was within its "legal rights." Diné activists called it "nuclear colonialism" and pointed out that Navajo first responders would be first exposed in the case of an accident.[184]

After exploring their legal options, the Navajo Nation signed an agreement with the company that includes public safety measures, payments to the Nation, and a promise to accept and process ten thousand tons of old abandoned uranium waste. Nation leaders said they got more than they would have without an agreement, but the decision was controversial within the community.[185] The fact that some environmentalists insist that nuclear would be safer this time does not reassure Diné opponents. For many, the legacy of environmentalists in Diné Bikéyah has put them in the same category as government and company officials, whose assurances are untrustworthy.

LOVING THE LAND AND HER PEOPLE

Founded as a hiking club in 1892, the Sierra Club gradually grew into a national organization that advocated for environmental protection. By the early 1960s, "the Club" was led by David Brower, just as Southwestern politicians and industry were devising plans to supply electricity to growing cities like Tucson, Phoenix, and Las Vegas. Behind closed doors, Brower agreed not to fight a hydroelectric dam in Glen Canyon, which was sacred to the Navajo, in exchange for the protection of Dinosaur Canyon, which was special to him. When two new hydroelectric dams were proposed for the Grand Canyon, the Sierra Club galvanized a massive grassroots and PR effort to protect Arizona's most popular and iconic vacation spot. Political leaders told Brower that if Grand Canyon was off the table, they needed an energy "plan B." Brower agreed not to oppose coal mining and generation on Black Mesa, which he described as "not very important country compared to Grand Canyon."[186]

The Sierra Club has apologized for this incident, which undermined the sovereignty of an Indigenous nation. I also heard another cautionary lesson in the story. Brower's description of Black Mesa as unimportant was based on his assessment of its value as a vacation destination, a common attitude in the primarily white branch of the environmental movement which he represented. This is very different than the understanding of nature described by Nicole Horseherder, founder of the organization, *Tó Nizhóní Ání* (TNA), which means "Sacred Water Speaks." In an article for the Club's national magazine, *Sierra*, Horseherder explained:

> Land is our connection to life. It tells us how to live. It tells us where to plant, where to graze sheep, where to hunt, where to gather medicine, and so forth. It is our mother, grandmother, great-grandmother. It nurtures and it whips. In old age, it cradles you and takes you back when you release your last breath. It plants grass over your grave and reminds you life goes on and you are merely a part of the earth and the cycle of life.[187]

In her description, land isn't a passive object to be admired, discovered, conquered, exploited, or protected. Instead, land teaches, nurtures, cradles, and humbles us. She is a living force with whom we are in relationship.

In *Braiding Sweetgrass*, Robin Wall Kimmerer asks her advanced botany students how many love the Earth, and every hand goes up. But when she asks how many believe that the Earth loves us back, they are quiet, surprised by this Indigenous perspective. This anecdote reveals a key difference between Indigenous teachings about loving the Earth, which many call Mother, and the view of nature I grew up with, where humans are, at best, admiring outsiders.

The US conservation movement developed amid this illusion of separation, with advocates willing to divide the Earth into places they loved enough to protect and places they were willing to sacrifice to deforestation and industrial pollution. This view intersected with the ways they divided and devalued humans. They often kept Indigenous, Black, Brown, Jewish, and working-class white people out of the places where they vacationed, such as the private reserves of the Adirondack Mountains in upstate New York. As widespread deforestation devastated the region, men like J. P. Morgan and William Rockefeller—founder of Exxon's predecessor, Standard Oil—bought up vast tracts of land for their own hunting and fishing.[188] Protected by legislation in the late nineteenth century, beaver, moose, and other species rebounded in those areas, even as Gowanus Canal is still dealing with the pollution of the early fossil fuel industry that was unchecked during the same decades.

I love hiking and camping in the regrown woods of the Adirondack Mountains and feel gratitude for their protection, even as I understand its limitations. A Cornell University scientist who studies pollution in Adirondack fish told me that the region's early protection has made it a good place to prove the impacts from outside the region. Mercury, released by burning coal, travels hundreds of miles from Midwestern smokestacks, hurting reproduction among loons, the Adirondack's most iconic bird. He said this type of research has influenced air pollution legislation, both in New York State and nationally. Science has confirmed what Indigenous wisdom always taught, that all life is connected.[189]

I NOTICED that Diné accounts of history consistently reference the people's reciprocal relationship with other species. This was clear in Barboncito's comment about being known by the snakes. Although sheep provided food and wool to make blankets, they were clearly much more than property in the traditional economy. When they slaughter sheep in the old way, Diné do so with reverence, facing the animal's head north and assuring the sheep that the herd will continue. Traditionally, people believe that sheep prayed to the gods for rain, helping the vegetation to grow. That is why a new baby's umbilical cord is buried in the sheep corral, so the sheep will look after the child.[190] Interestingly, new research shows that small doses of zinc may be able to heal damage done to the human body by uranium. Johnnye Lewis points out that zinc is found naturally in lamb and other favorite Navajo foods, like blue corn mush and piñon nuts.

When the Black Mesa Water Coalition envisioned an economy after coal, they realized that it needed to include sheep and other traditional ways of life. Younger activists set out to help those who make wool or Navajo rugs to get a fair price by cutting out the middlemen. In one event, Diné elders from Black Mesa and Big Mountain called for "resisters from near and far to converge" for "a continuous camp to defend Diné sovereignty and share traditional ways of life with the younger generation." Workshops on sheepshearing and dying wool with local herbs were framed as a form of protest.[191]

This spirit of empowerment is also shaping the work of the nonprofit Native Renewables, which trains local people, including former miners, to install modest units in remote areas to bring electricity to the thousands of Diné and Hopi who still live off-grid. With temperatures rising, having electricity to refrigerate food is especially important. "When a family member can manage their energy load and own their power, that's part of the self-determination and self-reliance that we are trying to attain, as a nation and as a people," said cofounder Wahleah Johns.[192] Building on her work with Black Mesa Water Coalition and Native Renewables, Johns later worked in the Biden administration Department of Energy, building equitable and sustainable energy access for Indigenous communities across the United States. Many are actively pursuing energy from the sun, which they point out has always been there for their people.

EQAT's green jobs campaign emphasized rooftop solar, which takes no new land and can be owned by families or community organizations, so the financial benefit goes to them rather than corporations. We were repeatedly told that rooftop solar was too expensive and unrealistic. Following the growth of Native Renewables over subsequent years, I was inspired that they were modeling a way of providing energy that encompassed multiple values: care of the land and the climate, care of energy recipients and workers, and sovereignty for their people.[193]

This approach to solar is very different than the profit-centric logic that has driven earlier environmental destruction, but that logic can also be applied to new "green" technologies. Large solar farms sometimes result in deforestation and habitat disruption. That's why Black Mesa Water Coalition wanted solar farms on the Navajo Nation to be built on land already disrupted by coal, an approach people are pursuing in many places with polluted land. It's good news for the climate that dropping prices and government incentives have spurred more renewable energy, as majorities in forty-seven US states now want to see a transition from fossil fuels to clean energy by at least 2050.[194] There is less of a consensus about whether we will make this transition in ways that move us toward a truly life-sustaining civilization for everyone, including other species.

COMPETING INTERESTS

On my last day in the area, the Navajo Nation and the US Park Service signed a historic agreement on the management of Canyon de Chelly, the only US national monument entirely on land governed by Indigenous people. Described as decades overdue, the 2018 agreement was intended as a first step toward a more comprehensive joint-management plan. The Nation had already taken over issuing various permits as well as managing the park lodge and concessions at the canyon. Sharing land management felt like another step toward a just transition.

While Indigenous teachings offer powerful guidance for how to live in balance with the Earth, today's Indigenous people include diverse perspectives and interests, just like any other group of people. At the signing ceremony, some canyon residents worried that fast-moving vehicles from tourism were dangerous for their children, while those who earned their living from tourism wanted to make sure the agreement didn't hurt business. The Navajo Nation government was focused on "economic development," especially in the face of the coal industry's decline. One Nation official talked about the need for jobs to keep people on the land that was secured for them by treaty. Meanwhile, the Park Service prioritized preserving resources, like the ancient Pueblo artifacts. It struck me that those who had gathered to witness the park agreement shared a deep commitment to the land, but they held competing conceptions of it: resource, place to visit, home, and sacred mother.

The agreement was signed by representatives of three bodies with a lot of history with each other: the Navajo Nation, the US Park Service (founded by John Muir, who also founded the Sierra Club), and the BIA (the agency that slaughtered Diné sheep in the 1930s). Everyone bowed their heads for an opening prayer in Diné, followed by a welcome in English. Speeches acknowledged that the collaboration they were forging was cutting edge and "about our children." They promised to continue listening to each other and the community. When they said they could take one question, an elder who was skeptical of the agreement stood and spoke at length in Diné, switching into English to point out that they had thanked each other, but nobody had thanked the people of the canyon. The officials responded in Diné, then English, thanking the people of the canyon.

Perhaps this is what progress looks like, I thought—inevitably messy. The encouraging thing was that, despite different interests, the groups in the room had agreed to keep talking and keep listening, something that had not happened in earlier periods. The region's superintendent of national parks acknowledged to me that when Canyon De Chelly opened as a park in 1931, the National Park Service did not conceive of its relationship with the Navajo as partnership. She was excited that they were moving in that direction now. I mentioned the concerned elder, who before the signing had stood outside with a cardboard sign protesting the agreement. "I don't blame people for being fearful," she said. "If you look at the whole history here, why wouldn't they be? So, I don't take it personally."

WHILE SOME people are working to break colonialist patterns, others are perpetuating them. Led by a French entrepreneur, a company called Nature and People First decided it would be lucrative to pump water to the top of Black Mesa when electricity prices were low and then let the water flow downhill to generate more electricity, likely for Phoenix and Tucson. Nicole Horseherder, others from TNA, and a group of allies successfully organized against the project's permit request. Although the company secured the support of one small Navajo community, two Navajo Nation departments submitted statements opposing the project to the Federal Energy Regulatory Commission (FERC). In a hopeful change from past practice, FERC denied the permits, stating, "The Commission will not issue preliminary permits for projects proposing to use Tribal lands if the Tribe on whose lands the project is to be located opposes the permit."[195]

The French entrepreneur, Denis Payre, wrote a 2024 opinion piece for the *Navajo Times* in which he complained about the FERC decision and blamed "NGOs opposing our projects." With a sense of entitled superiority reminiscent of past outsiders, he suggested that opponents were using emotion instead of logic. He claimed that his project would combat climate change, and aid economic development on the Navajo Nation. Most ominously, he asserted, "It's crucial to grasp that history has repeatedly demonstrated that not utilizing water resources can lead to losing access to them."[196] Payre's cynical claim that protecting water could diminish their right to it followed a 2023 US Supreme Court decision over access to water from the Colorado River, which is even more important to the Navajo Nation given how mining has polluted and depleted other water sources.

Running through several states, the Colorado is already over-tapped, especially by farmers growing feed for livestock. Old laws promised more water than the river can sustainably provide to states in the watershed. To rectify their omission from this agreement, the Navajo Nation argued that the 1868 treaty granted them a permanent home, which must include access to water, which the US government was obligated to provide. The Supreme Court ruled against the Navajo Nation, favoring the arguments made by the states of Arizona and Colorado, as well as the Biden administration. A White House statement after the court decision acknowledged that a ruling in favor of the Navajo might open a floodgate of claims from Indigenous people, something they were eager to avoid. After all, the Doctrine of Discovery remains a bedrock of US property law, which was reaffirmed in a 2005 ruling against the Oneida Indian Nation of New York. The majority opinion was written by Supreme Court Justice Ruth Bader Ginsburg, a reminder that colonization has been a bipartisan project.[197]

DEVELOPING SOLAR on the Navajo Nation and sharing Canyon de Chelly management could both be seen as win-win solutions that benefit Indigenous and non-Indigenous people alike. We need to pursue more of these options wherever we can, and resist any attempts to roll back such efforts. But as climate change exacerbates drought, the water issues in the Southwest could easily be seen as a zero-sum game with the interests of Indigenous people pitted against the interests of farmers, cities, and states. Likewise, nuclear energy could be good for the global climate while carrying catastrophic risks for those exposed to nuclear accidents. Even solar can be done in ways that facilitate just transition, or not.

If the illusion of separation prevails, policymakers will continue to make decisions based on what they see as their own narrow interests, with little regard for other people, let alone species. This is exactly the kind of thinking that created the crisis of the Earth in the first place. It's why a shift in consciousness that values justice as well as sustainability is necessary to support a truly just transition.

LESSONS FROM DINÉ BIKÉYAH AND BEYOND

So different from the moist green region I call home, the land of Diné Bikéyah gave me a longer-range perspective on the issues we are facing. On the eastern side of the Navajo Nation, the steep walls of Canyon de Chelly were shaped by water and wind. Formed hundreds of millions of years ago, the ribbons of sandstone swell and recede, leaving enough ledge for careful human feet. I zigzagged down the switchback path and across the floor to ancient Pueblo cliff dwellings. Continuously inhabited for thousands of years by different peoples, the canyon holds hundreds of archaeological sites. It also holds the memory of those who hid from the US Army before the Long Walk, and those who hid their sheep from slaughter a few generations later. The canyon reminded me not only of people's incredible capacity to survive catastrophic change but also of the patience of the Earth, which measures time in millennia—not in election cycles or even generations.

Diné speaker and writer Lyla June Johnstone shares Indigenous lessons for survival in her popular TEDx talk, "Three-thousand-year-old solutions to modern problems." Drawing from her doctoral research, she summarizes four of the land management strategies that native peoples have used around the world, including in her own dry homeland: "Work with nature. Expand habitat. Decenter humans. Design for perpetuity." Calling us to unite in courage and forgiveness, she says that Mother Earth needs us. "When we become her friend, her confidant, her ally, her partner in life, instead of her dominator, her 'superior' or her profiteer, we can transform dead systems to living ones." She concludes, "If our ancestors around the world proved this is possible, then it gives us hope that we can do it again."[198]

OVER THE years since my visit to the Navajo Nation, Indigenous ideas have become an increasingly important part of the consciousness shift that is slowly taking place. This will continue no matter who occupies the White House or Congress. While TED talks and books are helpful in spreading ideas, relationships are even more transformative, and more likely to lead to new behaviors, whether that means resisting a pipeline, installing solar panels, or returning land to its original stewards.[199] A small but growing movement, some call such land return "rematriation" because it affirms the role of women as land stewards, in contrast to the word *repatriate*, which connotes patriarchy.[200]

Lesili Haines, a Quaker from upstate New York, told me how she paddled with people from the Haudenosaunee Confederacy from Albany to New York City as part of the 400th anniversary of their peace treaty with the Dutch. Continuing to learn after that transformative trip led her and her wife to teach the difficult history of Indigenous boarding schools run by Quakers. During the same period, Michelle Schenandoah was supporting a group of Oneida women who longed for a place of their own to perform ceremonies on the land of their ancestors. Of the six nations of the Haudenosaunee, the Oneida had lost the most land. Michelle had been trained by a line of female Oneida leaders, as well as law school, to continue seeking the land's return. When Lesili heard Michelle speak about how the land had not heard Oneida songs for 200 years, she approached Michelle and said, "I have land in Oneida territory, and I'd like to give it to the Oneida women." Michelle was speechless.

It took time and relationship building to complete the transfer of thirty acres of recovering woodland that Lesili had inherited from her mother. Michelle felt the support of her female ancestors, especially her grandmother and great-grandmother who worked for decades to get their land returned despite their people being scattered from Wisconsin and Ontario to upstate New York. Michelle's account frames her collaboration with Lesili as one step in a long continuing journey that begins with the Doctrine of Discovery.[201] While Pope Francis renounced the principles of that doctrine in 2023, Michelle felt disrespected by other Vatican officials only a year earlier when she was part of a delegation of Indigenous and First Nations people who traveled to Rome to ask for the return of the children who died at Catholic boarding schools. For many Indigenous people, healing the land and the history are long-term, interconnected efforts.

Such connections were built on Long Island after Shinnecock women went looking for their traditional burial grounds so they could lay to rest children from the boarding school era whose bodies were belatedly returned. The Shinnecock were surprised at the welcome they received from Catholic nuns, who held title to that land. Out of this connection, the Sisters of St. Joseph began sharing land with the Shinnecock women who grow kelp in a hatchery there. Both groups hope that kelp can help clean up the carbon and nitrogen that is killing fish and shellfish in the bay where they live across from each other.[202] On a web page about their collaboration, one sister says that the Shinnecock women "awakened in us how to better live out our charism of unity and all inclusive love." The sisters say that they better understand what Pope Francis meant when he wrote, "Everything is connected."[203]

In 2015, Pope Francis published *Laudato Si': On Care for Our Common Home*, a groundbreaking encyclical that calls for an "ecological conversion" and challenges the hierarchical thinking at the root of environmental destruction. Of our relationship with the Earth, he writes, "We have come to see ourselves as her lords and masters, entitled to plunder her at will . . . We have forgotten that we ourselves are dust of the earth (cf. *Gen* 2:7); our very bodies are made up of her elements, we breathe her air and we receive life and refreshment from her waters." A true ecological approach "must integrate questions of justice in debates on the environment, so as to hear *both the cry of the earth and the cry of the poor*," he emphasizes. Calling this "integral ecology," Pope Francis roots these lessons in his own tradition, though articulating them in this way feels like part of the consciousness shift we so badly need.[204]

A few years after the release of *Laudato Si'*, I attended a program at a local Catholic university, where a priest in the audience said that his suburban parishioners didn't want to hear the encyclical's message. In recent years, my Catholic husband has seen much more engagement, especially from parishes and universities, which are trying to reduce their own contributions to climate change. On global Zoom calls, he noticed that North America is always underrepresented, with American church leaders more reluctant to move their money out of fossil fuel investments than leaders in other countries, even wealthier ones. His impression is supported by a study on the role of US bishops, which concluded that they "snuffed out the spark of *Laudato Si'*" by ignoring or undermining its message on climate change.[205] One notable exception is the Archdiocese of Chicago, which has committed tens of millions of dollars to supply all of its two thousand buildings with 100 percent green energy.[206]

As I witnessed on the Navajo Nation, the tension between worldviews is playing out within communities and nations—as well as between them.

PART IV

One World

CHAPTER EIGHT

ONENESS AND DIFFERENCE GLOBALLY

"Natural disasters unite," observed global climate advocate Harjeet Singh when we met in New Delhi, India's bustling capital. Jazz was playing above us at a coffee shop while outside cars, trucks, auto-rickshaws, bikes, cows, and pedestrians surged around each other amid constant beeping. Wearing a yellow-gold turban emblematic of his Sikhi faith, Harjeet said that offering free food was a standard Sikhi practice, but they had done much more of it after the raging floods in Southwest India two years before my 2019 visit. Hundreds of people were killed and hundreds of thousands lost their homes after extreme monsoons caused landslides. Amid the chaos, Sikhs fed Muslims, and Muslims offered Hindus refuge in their mosques.

"I think this is how humanity is going to behave," Harjeet predicted of future climate disasters, despite rising Hindu nationalism and prejudice against religious minorities in India. Smiling slightly through his thick black-and-gray beard, he asserted that helping each other was "a very natural instinct," one he hoped would grow to include human beings on the other side of the planet. This expansion of empathy struck me as another example of consciousness shift, the third aspect of the Great Turning.

I told Harjeet about the words that had inspired my research and travels: "The crisis of the Earth is saving us from our illusion of separation." He nodded, observing that the climate crisis was helping people around the world to realize that "it's one planet." During recent wildfires in the Amazon Rainforest, people in India were following the news coverage, thinking about the Amazon as the Earth's lungs. "So that's a good thing that has happened. We are far more connected. We can be a lot more united. We can be a lot more empathetic. This is a fantastic opportunity." He noted that developed countries have started experiencing more extreme weather themselves, and that has helped to build connection. "As it happens in any family, crisis brings you closer," he said. "You start ignoring smaller differences."

Harjeet recalled seeing a clip of an eighty-year-old man in the United Kingdom who glued himself to a train to pressure his government to act on climate. The man said he was committing civil disobedience for his grandchildren and also for people in poor countries who were disproportionately affected by climate chaos. Harjeet was encouraged. "We have been calling for climate justice for a very long time. In fact, sometimes we giggle when we now look at young people talking of 'Climate justice now!'" he chanted in imitation of a protest. He laughed and then said seriously that it is important that people understand what that phrase actually means.

THE CONCEPT of climate justice includes several layers. The first is about responsibility for the problem. Although Harjeet criticizes India's climate plans as insufficient when he is home, he defends India on the international stage, in the face of pressure from developed countries that built their wealth by burning fossil fuels. "The climate change we are seeing now is the result of those past emissions," he explained. China, now the world's largest emitter, has to bring down its emissions, too, but historically the US and EU contributed much more to the problem and have a responsibility to help developing countries to adopt greener pathways."[207]

Instead of support, Harjeet recalled 2013, when the US dragged India to the World Trade Organization (WTO), the arbiter of free trade rules. India's offense? It had started a major drive toward solar, and the Indian government offered financial incentives to Indian manufacturers to make the panels. "Now, what do you want?" he asked, throwing up his hands in exasperation. "You want us to go green, but only buy stuff from your companies? That's hypocrisy."[208] Still, the US won the case. Harjeet asserted that climate policies must account for development needs. Countries like Indonesia and Malaysia desperately need jobs for their people. Vietnam needs incentives to make its garment industry green, but those pushing global economic policies lack empathy for people in the Global South, he said.

Different terms are used for this empathy divide: North/South, developed/developing, or rich/poor. The lines themselves are fuzzy. Some countries that were considered developing thirty years ago, when global warming debates began, are now more industrialized, though they still haven't used fossil fuels for as long as countries like Great Britain, which also benefited from the colonization of what is now known as the Global South. In the nineteenth century, Britain deliberately suppressed local industries—like India's thriving garment and ship industries—and then forced colonized people to buy British-made products. As the formerly colonized began to win independence in the twentieth century, the Global North maintained its economic advantage through a set of policies collectively known as "neocolonialism."[209] This is part of why proponents of climate justice say that Northern countries should help countries in the Global South cover the cost of what are termed the "loss and damages" caused by climate chaos.

Poverty makes people especially vulnerable to extreme weather. Many millions of Indians work outdoors in agriculture, unable to afford air-conditioning amid rising heat. Floods are especially dangerous in India because so many people live in crowded, make-shift housing that is easily swept away. On city streets, women in saris squat selling vegetables, flowers, or spices to sustain their families, but such informal businesses have no insurance or government support when wiped out by a flood or cyclone. In fact, after flooding washed many homes away in Delhi in 2023, the government chased the survivors off the roads, telling them to "stay out of sight" as the government prepared to make the capitol look good for a visit from G20 leaders.

"I think sea level rise is going to be extremely dramatic," said Harjeet, who became a climate advocate after doing disaster relief in Bangladesh, where he met people whose homes were washed into the sea. He noted that India has a 7,500-km-long coast and will have to figure out how to relocate coastal populations, including fishermen who will need to find new ways to sustain their families. He was especially worried about food security for India's growing population as climate change increases drought and water scarcity.

I mentioned that a friend recently visited a village in Ethiopia, where her husband is from, and reported that people had stopped washing their dead because water is so scarce. Harjeet thought such examples helped to humanize the impacts of climate change, which will touch every aspect of life. "India is going to be extremely vulnerable, and we are not prepared," he concluded solemnly.

THE IMPACT OF WARMING

An esteemed journalist who has covered rural India for decades, Palagummi Sainath met with me in his Mumbai apartment. With white hair and deep-set eyes, Sainath has reported on India's most vulnerable—and has criticized the powerful—winning him scores of journalism awards as well as adversaries in government. As we sipped lemon water, he described the already high rates of suicide among Indian farmers, which I'd been reading about in the news. The causes of this crisis were complex, he said, but climate change was already exacerbating the dire struggles of Indian farmers.

Sainath explained that rice has been cultivated in India for seven thousand years. There used to be several thousand varieties—nature's way of ensuring that some crops survive in changing conditions. Then industrial agriculture pushed farmers to change traditional practices, making them more dependent on a few species, often with seeds controlled by multinational corporations like Monsanto. Even before that, colonialism pushed many farmers away from drought-resistant millet, a crop the British didn't value because they didn't recognize it, though it was the primary food source for people in South India. Such changes made Indians much more vulnerable to food insecurity amid rising temperatures and changing weather patterns.

India's growing water scarcity is not just from changing rain patterns. The Himalayan glaciers are melting, which will limit fresh water supply in a country that has 18 percent of the world's population but only 4 percent of its fresh water. Today, commercial cotton production in India—which uses huge amounts of water—has surpassed millet production. Sugarcane and rice also use colossal amounts of water. Exporting these crops amounts to what Sainath calls "virtual water export," increasing India's vulnerability to water shortage.

Sainath emphasized that the impacts of climate change vary across India's ecologically diverse regions. In the Himalayan Mountains, there are pastoralists who rely on yaks for milk, cheese, and meat as well as for the animals' hair and hides. "You know the yak is an amazing creature," he said. They are adept at living in the mountains, where temperatures get far below freezing at night. But yaks will die at temperatures that humans consider moderate. The pastoralists have moved their herds further up the mountains and started interbreeding yaks with cattle to make them more adapted to warmer weather. Still, they are likely to lose their livelihoods in the next twenty to thirty years, Sainath predicted. "Sheep are growing their coats thinner, so the quality of the cashmere is fading," he added about the region for which cashmere is named.

I said that in the United States concern for animals and concern for people are often treated as separate issues, but Indians seemed to be more aware of our interconnection. "You are not the only species inhabiting the Earth," Sainath replied, pointing out that our existence is often dependent on other species. "So, even if you approach it from the most selfish point of view, you still have to think bigger than yourself." In his decades of rural reporting, he had seen water scarcity impact both livestock and wildlife. "I am not an animal rights activist, but it makes sense to me that they have equal rights to water on this planet as any human being does. The difference between them and you is that you have the power to deny them water. They don't have a reciprocal power."

AFTER OUR meeting, I visited the online People's Archive of Rural India, founded by Sainath. One article describes huge declines in insects in India. One reason is that warming has led flowers to bloom earlier than they used to. "This means the insects don't get the food they need at the time they need it," explains Dr. Jayashree Ratnam, associate director of the National Centre for Biological Sciences, Bengaluru. When bee populations decline, it hurts the crops and fruit trees they pollinate, as well as the people who make their living collecting wild honey. This is happening around the world. Complex systems of interdependence are disintegrating, with effects we cannot fully predict. It's yet another example of the illusion of separation. Many people who live in cities don't think of themselves as dependent on insects, though farmers know better. Describing what is happening in India as a great drama, the article concludes: "The cast runs to trillions, of whom 1.3 billion are humans."[210]

A REPORT on the human cost of climate change was released while I was in Delhi. Produced by the Climate Impact Lab, it included a graph with a harrowing message: If the world continues on its current emissions course, by the end of the century, about 1.5 million Indians will die per year because of rising temperatures.[211] To a full amphitheater, Assistant Professor Amir Jina presented the results. With bright eyes and a matter-of-fact tone, he said that the temperature had recently reached 50.8°C in India (123.44°F). By 2100, instead of five extremely hot days per year, there would be an average of forty-two such days. Many people would perish in unrelenting heat waves. If humanity significantly reduces its current emissions—called "mitigation"—people will still die from the climate change put into motion by past emissions, but the deaths will be far fewer. "The benefits of mitigation are clear," he concluded.

Jina's talk was followed by a plenary by a government minister, who spoke at length in Hindi. The two young leaders of Extinction Rebellion (XR) India who had invited me to the symposium became agitated, and I asked what was wrong. They whispered that the minister focused only on adaptation—how India could adapt to rising temperatures—not mitigation, efforts to reduce greenhouse gases, so the world wouldn't get so hot. In their early twenties, the activists knew their generation would face the deadly consequences if dramatic changes were not made now. They wanted to hear how India could reduce its reliance on coal. But on the following panel, Indian policy experts predicted that India's energy use would more than double by 2040. Renewables, though growing, would not be able to meet the demand of such a large population increasing its standard of living.

University of Chicago economics professor Michael Greenstone asserted that the way to address the problem was through international negotiations. "Every country does a little bit more because if the US does something, it's going to benefit India, and then India will do a little bit more, and that will benefit the US." A few days later, I heard why getting every country to do more was so challenging.

THE COLLECTIVE ACTION PROBLEM

"If you invest a dollar in adaptation, you get 100 percent of that dollar in terms of gain to your country," explained Navroz Dubash in his book-lined office not far from parliament. If you invest a dollar in mitigation, the benefit is spread across the whole world. "So, a rational thing to do would be to invest in adaptation. Of course, that's true for everybody. So, how do you solve that collective action problem?"

Navroz had graciously fit me in before the launch of his latest book, *India in a Warming World*. He began our interview by recounting the evolution of international climate negotiations over his three decades of involvement. In the early years, India insisted that wealthy countries with high historic emissions should take responsibility for climate chaos, but those countries responded by portraying India as recalcitrant, the "bad boy of climate negotiations," Navroz explained. "That was kind of ironic because actually India was and remains a far more energy-efficient economy than many others." India ranks third in global emissions because its population is large and it is heavily reliant on coal, but its energy use per person is very low compared to the United States. India also started adopting solar early, when it was relatively expensive.

Understanding that India was itself deeply vulnerable to climate change, Navroz and others within India started pushing for a new narrative that emphasized "common but differentiated responsibility," where every country needs to do more, even though some need to do more than others.

To minimize the tension between climate mitigation and economic development, Navroz advocates a focus on "co-benefits," things like public transportation, which improve people's lives while reducing fossil fuel use. Energy-efficient lightbulbs are good for the purse as well as the climate, he pointed out. Adopting new, more efficient technology could give India long-run economic advantages. In fact, in the years after my visit, India became a leader in small electric vehicles like mopeds and three-wheel rickshaws, which are more affordable than electric cars and reduce India's infamous air pollution along with carbon emissions.[212] Navroz explained that such innovation was encouraged by the 2015 Paris Agreement, which allowed each country to choose its targets and how to meet them. Through it, the vast majority of countries acknowledged that we are one world, though with different domestic situations.

"India is no longer the bad boy, partly because we kind of got out of the way a little bit," Navroz observed. "And the rest of the world has gone rogue. So, suddenly, India is a leader among laggard countries." This felt like a polite reference to Donald Trump, who was in his first term as US president at the time of our conversation. In addition to announcing America's withdrawal from the climate accord, Trump had reversed many of Barack Obama's climate initiatives. A report in 2019 named India as one of the top three countries in the world for being on track with its Paris goals, while the US was "barely trying," in the bottom three along with Saudi Arabia and Russia.[213] Even worse, Navroz thought, was the effect the US withdrawal could have on the global spirit of collective action.

Although Joe Biden did not stop projects like Line 3 during his presidency, he did rejoin the Paris Agreement, which Donald Trump reversed as soon as he was inaugurated for a second time in 2025. As fourteen wildfires ravaged Los Angeles, he signed an executive order titled, "Putting America First."

COLLECTIVE ACTION on climate is not just about government policies. Several Indians I met criticized American consumption, from food to paper towels—"Five pieces for one tiny spill!" recalled one woman who had visited the States and was shocked by her host's wastefulness. I stayed with educated, professional Indians in both Mumbai and Delhi and noticed that my hosts walked more often than they used their cars, which like their homes were small by US standards. While our large water heaters run continuously, theirs are the size of a bucket, turned on only when needed, like electric teakettles. They hang their laundry to dry and use ceiling fans before resorting to air-conditioning. Indians also eat less processed food and much less meat, especially beef, which is forbidden to devout Hindus, who consider cows sacred. Although I turn tat-tered clothes into rags and vegetable waste into soup broth, I still consume much more than most Indians.

Navroz noted that the average carbon footprint of a person in the United States was three times the global average, which Indians were still far, far below. "If you have a finite amount of carbon to burn, and you are interested in maximizing global welfare, who would you give that carbon to?" he asked. Navroz asserted you should give it to the world's poorest, such as the 600 million women around the world who cook over fires of dung and wood. There was a push in India to help such women acquire gas stoves.[214] That would increase emissions modestly, but it would also protect those women from inhaling toxic smoke, save them time collecting fuel, and benefit the soil, which is deprived of nutrients when biomass is burned. "That's an area where to me the answer is clear," Navroz said. "Those people should not be carrying the carbon monkey on their back."

Such women do the least to cause climate change, but they are among the most vulnerable to its effects. This sunk in several years ago when I participated in a workshop on gender inequality. As the facilitator gave global statistics on unequal literacy and nutri-tion by gender, I realized with a jolt that most of these disparities will get worse with climate disruption. Drought will force many women and girls to walk farther to collect water, prompting even more families to keep their girls home from school. Fed last in many places, women will be disproportionately affected by rising hunger, along with queer, trans, and nonbinary people who are marginalized in most societies. Increasing natural disasters will force more people into refugee camps, increasing incidents of sexual assault. Gender inequality is another aspect of climate injustice.

Although Navroz acknowledged that Indian politicians should not "hide behind the poor," as Greenpeace once accused,[215] he asserted that you need energy to decrease child mortality and other measures of extreme poverty, and renewables can't yet meet the needs of so many people. Referring to the Paris Agreement's aspiration to keep global temperature rise to 1.5°C, he observed, "You clearly can't achieve 1.5 without major sacrifices. Should we be the ones to sacrifice?"

While outright climate denial is a partisan issue in the United States, deflecting blame for climate change is a bipartisan problem. Even in the more progressive spaces where I give speeches, I am often asked, "What about population growth?" The implication (intended or not) is that higher birth rates in countries in the Global South are a problem—not the longer and more carbon-intensive lives of people in the Global North.[216] The temptation to point fingers exacerbates the collective action problem, which is hurting us all.

THE TRUE COST OF COAL

While it is unfair for Americans to blame Indians for climate change, Indian activists point out that it's also unfair for India's poorest to suffer for the profit of the coal industry. "We always say that coal is cheap, but why is it cheap?" asked Priya Pillai, an intense lawyer and community organizer with long, pulled-back hair. She explained that it's cheap because the people who are displaced by coal production are not fairly compensated for their land, their livelihoods, or the health impact of living near coal production.[217]

Poor people are also disproportionately hurt by the suffocating urban air pollution that results from burning coal. "If my son has asthma, I am able to buy him an N-95 mask or take him to a doctor. But the ragpickers, the homeless, the rickshaw pullers, the hawkers, the urban poor who live in slums, they are most impacted by it and the least resilient. They end up spending a large amount of their income on health impacts due to air pollution," Priya explained. I was already tired and coughing from my time in Delhi and couldn't imagine working outside all day.

I'd come to India to understand how the division between the Global North and the Global South got in the way of climate solutions, but of course, there were also huge disparities within the country. India's caste system is a hereditary form of separation that still keeps millions locked in poverty. The British exploited this ancient stratification during the colonial era by recruiting India's upper caste to represent the empire's interests. In addition to caste, there were other lines of separation, including gender, religion, and geography. I looked up Priya because she spent four years building a successful partnership with Adivasi—often translated as India's "Indigenous" because their deep relationship to the land predated the arrival of other populations that are part of today's India.

IN CENTRAL INDIA, an estimated fifty thousand Adivasi were sustained by the Mahan Forest, where they harvest roots, herbs, fruits, and flowers with medicinal properties. The ancient forest is also home to elephant, leopard, and many other species. It was supposed to be off-limits to destructive coal development, but two companies acquired clearance to mine.[218] At the time, Priya was working for Greenpeace India, which offered to support the villagers, who initially didn't know that they had ancestral rights to the forest that were protected under the law.

Priya noted that climate activists tend to be urban and middle class. They talk about 1.5° and the fact that forests are carbon sinks, places where the Earth naturally stores carbon. "For people on the ground, it's their livelihood," she stressed. "It's their land. It's their forests." The people have a spiritual connection to the place where their ancestors worshipped for generations.

Priya emphasized that it didn't work to describe climate change as something "up there," gesturing skyward. "Maybe the people on the ground will not articulate it as climate change because there are more pressing issues that they're dealing with, but ultimately the battle is the same because it's interconnected." I remembered Anne Rolfes saying the same thing about Black communities in Cancer Alley.

Priya described the slow process of building trust with the villagers, who didn't want to talk to her when she first arrived. She spent two years building relationships, more time than results-oriented NGOs are usually willing to spend. "For me as a city-bred person, it was a journey of relearning," she recalled. As a lawyer, she helped the people to understand their legal rights and shared information only published in English. To build grassroots power, she had to wait until the people felt ready to take ownership over the campaign. Once that happened, they were able to mobilize thousands of people in a region where villages were small and dispersed. She stressed that their successful collaboration did not come from Greenpeace throwing money at the situation. When they needed funds, they asked the people, who would give five or ten hard-earned rupees each, comparable to US pennies. "But everybody had a stake, and that stake is what mattered." People felt they owned the movement.

They faced many obstacles as the campaign grew. Local leaders, in bed with the coal company, forged documents and attacked Greenpeace, which made people more sympathetic to the global organization. On one occasion, mining company employees started marking trees to be killed, and during their lunch break community members collected company equipment and delivered it to the police. That night, four campaign leaders were arrested, one held for twenty-eight days. Three rounds of arrests and threats of violence strengthened the people's resolve, as it often does.[219]

In a major turning point, Priya made international news in 2015 when the Indian government prevented her from boarding a plane to the UK where she was going to speak.[220] The government accused her of acting against the national interest, but the suppression backfired, giving Priya an even larger audience. The courts ruled that she had been wrongly denied travel. Soon after the Ministry of Environment, Forest and Climate Change ruled that the forest would remain protected, over the objections of the Ministry of Coal. After the announcement, Bechanlal Shah, the villager who had spent twenty-eight days in jail a year earlier, told a reporter, "Celebrations started as soon as we heard the news! The government has finally accepted that this forest, which gives thousands of us so much, must not be destroyed for the profit of a few."[221]

Shortly after the victory, Prime Minister Narendra Modi marked his first year in power by blocking international funding for nine thousand NGOs, citing Greenpeace as a threat to economic growth fueled by coal, one of the issues he had campaigned on.[222]

ONE ANECDOTE that Priya shared stayed with me long after our meeting. She had brought some Adivasi villagers to Delhi during the forest-protection campaign so they could meet a government minister and experience the system they were up against. When they visited a mall for the first time, they wondered why people rode the "stairs that move up" and asked if they used energy. She said yes and explained that escalators were more convenient. The villagers were shocked that city people were so lazy and heartless. "Do they even realize we are being displaced for this energy?" one man asked. Recalling the story, Priya acknowledged that as a city person, she was part of the energy-intensive system that catered to the wealthy and middle class. "We really don't care about the true cost of that coal," she said sadly.

A CRISIS IN EMPATHY

I believe that we all have a stake in addressing the climate crisis, but many of us don't yet feel the true cost personally. To secure a stable climate for everyone, and help each other cope with the effects already put in motion, we need to cultivate empathy for those already suffering. Harjeet Singh was heartened by people feeding those of different faiths after raging floods in Southern India. But his optimism dimmed when he shared that some Indians are worried about climate refugees coming from neighboring Bangladesh, which is predominantly Muslim and even more vulnerable to flooding than majority-Hindu India. Harjeet recalled an ancient Indian principle, *Vasudhaiva Kutumbakam*, which translated from the Sanskrit means, "The world is like a family." He continued, "The world is One. It's not even a village. It's one family. That expression is there in our scriptures. So, why can't we execute that vision? Why can't we feel . . . ?" His voice trailed off with a sad chuckle.

Empathy is defined as the ability to understand or share the feelings of another, so "Why can't we feel?" is the right question. I did not grow up in a family where feelings were encouraged, especially grief or anger. I did grow up with stories about the nineteenth century famine that the Irish called "the Great Hunger."

My mother blamed the English, meaning the wealthy landlords who continued to export food as the peasants around them starved. Later, I learned that working-class English people collected money for Irish famine relief. So did the Choctaw only sixteen years after their brutal forced removal from the Deep South to west of the Mississippi River. In Cork, Ireland, there is a monument to Choctaw generosity called Kindred Spirits. It consists of a circle of nine giant stainless-steel feathers that form an empty bowl. It is a beautiful reminder that we can always choose connection, especially when we recognize that our troubles are not separate from other people's.

Knowing that my own ancestors were displaced by famine helped to puncture my illusion of separation when I learned how climate change would make famines more frequent and severe in my lifetime. This realization hit when my best friend from Botswana called me out of the blue. Twenty-five years after I taught there in the Peace Corps, I followed Tswana custom and asked about the weather. She said it was extremely hot, prompting me to research climate change in Botswana after we hung up. I learned that the rains had become so unpredictable that farmers didn't know when to plant. Worries about food insecurity were growing, and the government was encouraging farmers to switch back to precolonial crops, like sorghum and millet, which were more drought resistant than the corn the British had introduced. If top-emitting countries didn't reign in their greenhouse gas emissions, climate-related deaths in Africa could reach unfathomable numbers.[223]

Wanting to share my increased sense of urgency, I wrote about African vulnerability to climate change,[224] but I found that message did not motivate many Americans. Once at my local library, a scientist presented the usual scary graphs predicting global temperature rise. His tone seemed detached, as if his statistics would have no human impact, which I asked him about during the Q and A. "Oh, yes, climate change will kill a lot of people, but they are mostly in the Third World," he said. He instantly cringed at his own words. Or maybe he was responding to the look of horror on my face, which I followed with a rant about the need to see those people as human beings. The audience sat frozen in awkward silence.

I believe that racism contributes to American apathy toward people in the Global South, exacerbated by the fact that those countries are rarely covered in our history classes or on our news. Yet, that's not the whole story. Feeling the devastating impacts of climate chaos is also hard because its worst effects will be in the future and the scale challenges the human imagination. It doesn't help that many people still think of polar bears when they think of climate victims. When we do hear the horrific human death predictions, we may feel guilty or powerless.

None of these are uniquely American challenges. At the climate symposium I attended in Delhi, a television journalist, who served as panel moderator, began by acknowledging that the forecast was grim. He glanced around the amphitheater awkwardly, as if to see if he was the only one shocked to hear that a million and a half Indians per year were likely to die from rising temperatures. Other than the young activists next to me, most in the professional audience did not reveal much emotion.

Joanna Macy urges people to grieve for the world as part of an approach she calls "the work to reconnect." She points out that grief for the Earth and other human beings is born of love, and anesthetizing ourselves to those feelings dulls our creativity and ability to take action.[225] I've noticed in activist spaces a growing interest in somatic exercises to get people in touch with their bodies, which can help them access and integrate their feelings.

IF I had not found an effective and spiritually grounded group to take action with, learning about the human cost of climate change might have led me to numbness or despair. Fortunately, in recent years, both action and empathy for people in other countries have grown among grassroots climate activists, like the eighty-year-old man in the United Kingdom who impressed Harjeet by gluing himself to a train, not just for his own grandchildren but also for people from poor countries. Unfortunately, it has been harder to find empathy among the world's most powerful.

Since the 1992 Rio Earth Summit, poorer countries have been asking for financial help to deal with the consequences of climate change. Pledges were made in 2009 and then as part of the 2015 Paris Climate Agreement. Wealthy countries promised that starting in 2020 they would raise $100 billion per year for developing countries. When 2020 arrived, that number was declared unrealistic. Meanwhile, the financial burden of climate chaos was increasingly felt on every continent. In 2022, the unequal impact was exemplified by the image of Pakistani women wading through biblical-scale floodwaters with cooking pots on their heads and children in their arms. The catastrophe displaced nearly 8 million people and killed several thousand while wiping out crops, livestock, roads, and homes in one-third of Pakistan.

A few months after the floods, the world's climate negotiators met in Egypt, and countries from the Global South united to push the issue of loss and damages to the center of the agenda. When the United States and other wealthy countries resisted, *The New York Times* cited "a crisis in empathy" as a key barrier to action.[226] After two weeks of acrimony, rich countries gave in and agreed to create a loss and damage fund to help poorer countries stricken by climate-related disasters. While seen as an important step, many commentators lamented that so much time and goodwill was lost fighting over the money that the conference failed to make progress on the issue of mitigation, rejecting a proposal by India that would have committed the world to reducing all fossil fuel use (not just coal, as countries that primarily used oil or gas preferred).

When Donald Trump returned to power in 2025, he withdrew the United States from the loss and damage fund, again under the banner of "Putting America First." The US had only pledged $17.5 million, much less than France or Italy.[227] The amount was a tiny fraction of the $30 billion in economic losses that Pakistan suffered in the 2022 floods.[228] Our pledged amount was also less than the United States gives to fossil fuel companies in tax breaks.

CONNECTING OUR WALK AND OUR TALK

I noticed that many Indian climate advocates accused the Global North of hypocrisy, including Harjeet Signh and Navroz Dubash after the conflict at the 2022 climate negotiations. They were especially frustrated when US actions didn't match US rhetoric. It struck me that hypocrisy is another form of separation—separation between what we say and what we do. I think of integrity as integrating our walk and our talk—and being willing to admit where we miss the mark. One of my favorite lines from the Gospel is Jesus's question to his apostles: "Why do you see the speck of sawdust in your brother's eye and pay no attention to the plank in your own?" Of course, most of us would rather ask other people that question than face it ourselves.

The extreme wealth gap in India helped me notice how my actions sometimes felt disconnected from my values. For example, when the subway or train wasn't convenient, I often used Uber, even after I learned that the global company had lowered the rates

Indian drivers could charge. Many were barely breaking even after paying for gas, tolls, and sometimes for the vehicle's rental. Feeling guilty, I decided to increase my tips, aware that didn't fix the structural injustice. I noticed that I preferred giving a generous tip on a cheap fare rather than negotiating a higher rate with independent drivers, whom I was told would overcharge me for being a foreigner. Even when it cost me more money, feeling generous boosted my ego, while feeling gullible bruised it. Uber also saved me from mispronouncing my destination to taxi drivers, which once got me lost. Fundamentally, Uber made me feel more in control, which felt like a spiritual issue at its root.

I gave up this control one day when I took a Mumbai train and disembarked in a neighborhood where street names were hard to find, even with an iPhone map. I stopped a man on the street for help, and he did not speak English, but accepted my phone when I called the person whom I was meeting. After hearing where my appointment was, the man kindly walked me the remaining blocks to my destination. Such moments of vulnerability helped me to appreciate Indian hospitality, which I later heard was based on the belief that "guests are like gods." Only later did I wonder how many such encounters I missed by trusting an app more than a stranger.

THE DESIRES for convenience, control, and ego-boosting status are part of what is driving the climate crisis, especially as American forms of consumption are exported to elites around the world. While energy access can improve the quality of life of poorer people, there is much evidence that an endless quest for bigger houses, overflowing wardrobes, and fuller bellies is not actually good for any of us. It's not just that eating more meat and processed food is unhealthy. Competition to prove one's worth through the latest fashion or gadget produces more anxiety than joy. Private planes, pools, or even cars are more convenient than public facilities, but they separate us from other people, while using more of the Earth's resources. In Mumbai, a luxury apartment building was designed to include a private pool on the porch of every single apartment, despite India's dire water scarcity. Sainath's public criticism of the project got the plans amended, but he told me that the government felt threatened by his journalism.

While it's right for Indians to question their country's direction, I try to focus on "the plank" in my own eye, as Jesus suggested. I know that decreasing my own consumption is not likely to help anyone in India, but I still try to be mindful of my choices. My husband and I buy our energy from green sources and use ceiling fans, reserving air-conditioning for extremely hot days, which of course are becoming more frequent. We share one hybrid-electric car, augmented by public transportation. For as many actions as I take, there are at least as many that I fail to do, such as giving up air travel. When I am successful at making changes, it is not because I think of them as effective ways of changing the world, but as spiritual practices meant to change myself, to loosen my attachment to comfort and convenience.

INTEGRITY IS an important principle for Quakers, many of whom gave up eating cane sugar in the nineteenth century because it was harvested by enslaved people. Today, it is impossible to be part of the global economy and not consume anything tainted by exploitation. Trying to be pure is its own ego trap. Although most climate activists I know reject the focus on individual purity, which has been encouraged by the fossil fuel industry, many quietly acknowledge that we could be doing more to reduce our own consumption. For me, the key question is what I feel Spirit is leading me to do. I feel called to a type of organizing that requires time, internet access, and the ability to travel. But I also feel clear that I need to be willing to make some personal sacrifices in order to speak with integrity when I challenge CEOs to do better. This is very different than saying that I have to be perfect before I criticize the corporations and governments that have constricted all of our options. If that were the case, none of us would ever take on the powerful.

In India, I thought about these contradictions every time I passed a statue of Mohandas Gandhi. In my twenties, I'd read Gandhi's autobiography and been impressed with his commitment to self-sacrifice. Although trained as a lawyer, he wore simple, homespun clothing to show solidarity with the poor and support India's economic independence from Britain. He also fasted, a spiritual practice I've never mastered. As I learned about the tradition of nonviolent direct action, I came to know Gandhi as a brilliant strategist, who turned the willingness to sacrifice into an effective mass political tool. The Indian movement for independence, in which he played a key role, is a stunning example of people coming together across lines of caste, gender, and religion for a common goal. In recent years, I've heard growing critiques of Gandhi, especially for his attitudes on gender, caste, and race, which were not as egalitarian as many Westerners assume. In India, I met people quick to point out the ways he did not walk his own talk, while others still hailed him as an inspiring example.

Gandhi's complex legacy offers many lessons, some strategic, like the need for nonviolent resistance alongside efforts to build alternative institutions. Other lessons are spiritual, like the reminder than none of us are perfect. We can't predict how history will judge our choices. This is no excuse for real moral failings. It's simply to remember what Daniel Hunter said, that if perfection is the bar to entry, we'll have a pretty quiet movement. Likewise, if we cancel every historical leader who fails the test of hindsight, we will lose lessons we sorely need for today. That, of course, is what the powerful want.

FINDING CONNECTION IN IMPERFECTION

The city of Varanasi exemplifies a spiritual tradition where God is found in an imperfect world, not apart from it. I arrived there worn down from Delhi's thick air, but eager to visit the Ganges River, known as the goddess Ganga Ma to Hindus. Believed to be a Hindu's direct route to Nirvana, many families bring their loved ones to Varanasi for cremation. When I reached the river, several men stepped forward to offer me a ride in their boats. Weary of being solicited, I replied, "I'm here to pray," and the men respectfully backed away. I stood alone in prayerful silence as tourists and Hindu pilgrims walked past. For the first time on my journey, I felt lonely.

The Ganges headwaters are in the Himalayan Mountains near the border with Tibet. By the time she reached Varanasi, India's most sacred river was rife with pollution, though Modi's Hindu-nationalist government had pledged to clean her up. One problem was that many families couldn't afford enough wood to burn their loved one's body completely, so there were pieces of human remains left by many cremations. Like Gowanus Canal, the river contained raw sewage, as well as waste from tanneries. While the old New York tanning industry had once stripped the Adirondacks of hemlock trees and polluted the waterways of the city, most leather today is tanned not with bark but with chemicals that are linked to high rates of liver and pancreatic cancer. Despite these dangers, many people bathe in Ganga every day, believing that the holy waters can wash away sin.[229]

I joined thousands of others for the evening Aarti, the daily sacred ceremony in honor of Ganga. On raised platforms facing the river, seven Hindu priests repeatedly bowed and raised tiered lamps into the dark sky. The sweet scent of incense mixed with the stench of sewage. The scene behind the crowd reminded me of a carnival—hawkers selling beads, carnations, and something that looked like pink cotton candy. I noticed an old Indian man with a wild beard wearing a loincloth, his body rubbed white. He demanded a donation from a tourist after demonstrating his ability to hang upside down from a nearby bamboo scaffold. I noticed a young Indian man also watching this scene with amusement, and our eyes met. We both laughed, and I no longer felt alone.

As the ritual continued, I reflected on all the ways I could feel separate from the people around me, and all the ways we were not so different. I'd worked as a door-to-door fundraiser, so I could empathize with the hawkers and tour guides trying to get me to listen to their pitch. In my mind, I had divided the crowd into tourists and worshippers, but I was both, seeking something more lasting than a selfie. I remembered a wealthy Indian woman I had met in Delhi, who told me, "We are all one," even as she insulted her servants and the poorer Indians beyond her lovely home. I had judged her for being judgmental, doing the very thing I'd criticized her for. Judgment is just another way of separating ourselves, and I could feel it seeping away by the Ganges. I could see myself in the woman I'd criticized, and also that of the Divine in her, just as I see both myself and the Divine in the performing yogi.

Near the end of the ceremony, a man thrust a large flat collection plate into different parts of the crowd. Some pretended not to see it, though most threw down money and received a hasty red smudge mid-forehead. I asked the Indian man next to me if it was all right for a non-Hindu to receive the red mark. "Of course!" he exclaimed. Later I was told that Hindus don't draw those lines because we are all One.

I WAS so moved by the ceremony that I decided I would put my legs in the Ganges the next morning. The doctor who'd given me my overseas immunizations had said that it was probably fine, so long as I had no cuts on my skin for the cholera or other health hazards to seep in. As I walked from the ceremony back to my hotel, I felt a surprising slash of mild pain on the top of my foot just over where my shoe ended. My phone flashlight showed a thin red cut. I could find no logical explanation for the wound. I took it as a sign that I wasn't meant to put my legs in the river when I had a cut and a depleted immune system. I felt like I was being cared for by Spirit, which doesn't remove the hazards of the world but can guide us through them. It was a reminder of all the ways I had felt guided and supported on this journey.

When I left Varanasi a few days later, I felt renewed on some deep level. The city's grittiness only deepened its spiritual effect on me. It's not a place where people worship a God of the sky. They worship a God come to Earth in the form of a river, even if filled with bones and feces and toxic pollution. We need more of this kind of spirituality, daily and in the muck, not just inside quiet, protected walls. If we are going to heal our relationship with the Earth, Spirit, and each other, we have to acknowledge and be in the muckiness of the world, while also celebrating—even worshipping—its tragic beauty.

LESSONS FROM INDIA

Although I was warned not to buy into the Western stereotype of India as a uniquely spiritual country, I was frequently reminded of its spiritual teachings. Small shrines were everywhere, often decorated with colorful flowers. During the festival of Diwali, even a home's threshold is colorfully decorated. When I asked my hosts Nandini and Tim if we would be going to temple for the festival of lights, they said that you don't need to go to a temple or church or mosque to find God because God is in everything. God isn't separate. This affirmed my sense that teachings from diverse spiritual traditions can aid us in addressing the crisis of the Earth. At the same time, the contradictions of India helped me acknowledge that fully living our deepest beliefs is a human challenge, not just an American or Christian one. What the Anishinaabe prophecy called a choice between the scorched path and the green one was here, too.

One day, as we walked around southern Mumbai, Nandini and Tim showed me a large, sacred tree near their apartment. There were stone platforms around its wide base, with statues of several Hindu deities—each a different manifestation of the divine Oneness. Incense and a garland of bright orange marigolds adorned the altars. Tim told me that even if a sacred tree is falling over, people won't cut it; there's so much respect for the tree. Having heard about all the Indian forests razed for coal mining, I quipped, "Unless there's coal under it. Then the tree gets cut." He nodded. Of course, big companies drove that destruction, often with support from government.

There were many Indians pulling in the other direction. I met one former city planner who was helping villagers to greatly improve their incomes by growing mango, papaya, custard apple, and other fruit trees, which would in turn help the soil and attract more rain. I met people who were aiding the regrowth of the forest in a region that had been felled by the British during the colonial period. They were healing the harm to the Earth and teaching younger generations about the medicinal uses of native plants. After my trip, I heard about people planting almond trees along the banks of the Ganges, which has brought back the birds while helping to hold the soil in place during monsoons. Others in Varanasi are collecting devotional flowers and recycling them into incense. New electric crematoria are honoring an ancient belief in a way that reduces waste.[230]

JUST AS the crisis of the Earth has the potential to help us overcome our illusion of separation from other species and other communities, it also has the potential to help us transcend the boundaries of nation-states. In fact, it will need to, since no one country can solve the climate crisis on its own. Clearly, we are not there, yet. While the expansion of a global climate movement is inspiring and necessary, national leaders continue to approach climate negotiations as a zero-sum game, with the United States a frequent obstacle to justice, regardless of which party is in power. Nationalism and religious bigotry have increased in many countries at exactly the point in human history when we most need to orient toward our common humanity—not just to prevent catastrophe but also to cope with it. The world keeps giving us opportunities to learn this lesson.

CHAPTER NINE

LESSONS FROM THE
COVID-19 PANDEMIC

When actor Tom Hanks caught the COVID-19 virus early in the pandemic, it was a reminder that anyone could be affected, regardless of privilege. Yet, when celebrities tweeted "We're all in this together" from their mansions, the message fell flat. They were speaking of our oneness without acknowledging that many people had jobs or housing situations that didn't allow them to isolate. A host of factors related to racial and class inequality made the pandemic more lethal in some communities, including many I had visited in my travels. Like climate chaos, the suffering caused by the pandemic was shared, but unequal.

The pandemic also demonstrated that policies and ideas matter, especially in a crisis. While governments around the world scrambled to figure out effective policies, Americans who mistrusted government were more likely to refuse to wear masks, get tested, or get vaccinated. I couldn't help but wonder if this was another legacy of the campaign to undermine trust in government. Research indicates that both noncompliance and social inequality contributed to the United States having one of the world's highest death rates, especially for a developed country.[231] Our experience of COVID-19, in turn, seemed to exacerbate some people's mistrust of government.

At the same time, the pandemic brought together scientists from around the world in an unprecedented way, leading to the development of a vaccine in record time. More people became aware that their fate was inseparable from those in other countries. For some, the crisis brought out a sweet deepening of community. When our two adult children came to stay with my husband and me, our daughter recruited us to help sew masks for strangers (until we learned that the homemade ones didn't work well). People in our neighborhood masked both for themselves and for the protection of others, at least in the early months. People donated food for those hurt economically, and other forms of mutual aid flourished. I hosted a range of online events, from an impromptu reunion of college friends to a daily rosary recitation during my mother-in-law's final weeks. Family who couldn't be together in person, because of the pandemic or distance, lingered on the calls catching up.

For many of us who had enough food and social connection, the pandemic provided glimpses of a simpler way of life. I stopped driving to the gym and started walking new routes around my neighborhood. I became acquainted with a huge gingko that must be a few hundred years old, a tree I had somehow never noticed from the car. I hosted a Facebook Live discussion with Sherri Mitchell, Penobscot author of *Sacred Instructions*, who reminded me that the Earth provides the medicines we need. She suggested that people pay special attention to any new plants that appeared during the pandemic, and I was amazed to realize that my raised bed was suddenly full of purslane. I had assumed it was a new weed, but with Sherri's encouragement learned it is not only edible, but believed to strengthen the immune system and the heart. I threw the succulent into salads, remembering Robin Wall Kimmerer's assertion that the Earth loves us back.

Some said that Covid itself was a message from the Earth, since deforestation and other habitat loss was pushing threatened species into greater contact with humans, increasing the emergence of viruses like COVID-19. Other stories from the natural world were more hopeful. Without the prying eyes of zoo visitors, pandas in both Hong Kong and South Korea produced long-hoped-for cubs. Globally, fewer animals were killed by planes, cars, and ships. Carbon emissions and air pollution were greatly reduced by people staying home, at least temporarily.

For the climate movement, however, the pandemic interrupted the momentum coming out of the youth-led climate strike that mobilized several million people across 156 countries. Activists who couldn't meet in person struggled to adapt, often amid social justice protests, such as those around race in the United States, and those for farmers' rights in India. To help climate activists find new footing, Daniel Hunter, his colleagues at 350.org, and a few other groups organized an online gathering for people around the world. After more than a year of living under pandemic restrictions, I was eager to connect with fellow climate activists, even though it was through my laptop and an interpreter.

FROM FIJI TO PHILADELPHIA

The Global Just Recovery Gathering included seven thousand people from 151 countries. Over three days in April 2021, two hundred sessions included interactive skills workshops, speaker panels, grassroots storytelling, and cultural offerings, like the Fijian welcome that opened the event at 3 a.m. Philadelphia time. The traditional ceremony emphasized gratitude and connection. Knowing that rising sea levels have already created ghost towns along Fiji's coasts, I wept quietly at the bittersweetness of the moment. My tears continued to flow as I watched a group of colorfully dressed

Pacific Islanders sing a harmonic and hopeful song about their relationship with the sea, while swaying in front of a majestic tree that reminded me of an old New Orleans live oak.

During the opening panel, Canadian author and activist Naomi Klein said that the pandemic had given everyone on the planet an embodied experience of rapid societal change, disproving the usual doomsaying excuse that change was impossible. The challenge now was for us to assert a bold vision for transformation that was better than the status quo. A society oriented around care of the Earth and each other might include free public transportation, which would reduce pollution, traffic, and climate-warming emissions, while saving people money. This type of "beautiful vision," she asserted, was much more appealing to most people than dimming the sun or other misguided interventions that geoengineers were proposing. "Just recovery," the Gathering theme, was an opportunity to reorient around a different set of values, not just profit and consumption. During the pandemic, she reminded us, "People didn't miss shopping. They missed each other."

Feminist leader Hakima Abbas of Kenya built on Klein's point. "During this pandemic, I think we've all seen the centrality of care in a thriving society. So, if we can imagine an economy centered around care rather than production, it would be a completely different logic," she said, noting the disproportionate role women play in caregiving. Acclaimed Indian writer Amitav Ghosh observed that young people were already experimenting with more caring ways of living, especially at activist encampments like Occupy and Standing Rock. Noting that authoritarian leaders like Donald Trump and Jair Bolsonaro of Brazil had not been able to achieve all the environmental rollbacks they wanted, he asserted: "The lesson of this period is that activism works." Although many speakers used terms like "solidarity" or "unity," it was a unity that acknowledged oneness and difference, including the fact that some countries were using violence and draconian legal measures against people protecting their communities from industry.

I WAS honored to colead a workshop on nonviolent direct action with Kumi Naidoo, the former head of both Greenpeace International and Amnesty International. Kumi first became an activist as a Black teenager under the brutal system of apartheid in South Africa. He noted, "We are acting out of love—for our children, for other species, and for people who look different than us." At the same time, he acknowledged, we often struggle to talk about the ways we are divided by race and the extremely unfair fact that the people paying the highest and most brutal price of climate change are the people in the world who did the least to cause it. He encouraged participants to have difficult dialogue about these issues "with optimism, love, and a sense that the purpose of having these conversations and debates is about moving forward." Coming from a country where music, theater, and other art forms were integral to the struggle against apartheid, Kumi said that we needed to change the culture in order to make the bigger changes we sought.

This approach was woven into the online conference, which included music, poetry, and other forms of cultural sharing. I later asked Daniel why they prioritized this, sharing how moved I was by the Fijian welcome ceremony. He explained that they didn't want a series of talking heads, like so many conferences. They wanted to touch people's souls, especially after a year of COVID-19 losses, lockdowns, and stalled plans. Music could do that across the barriers of language. "Like you, there were pieces of poetry, pieces of music that just brought tears to me," he recalled. "That meant that when I heard a panelist speak, I was listening with different ears. I wasn't listening with my rushed, analytical, Covid-drenched self. I was listening with ears that had been reattuned to ask 'What's the truth in this?'"

The Gathering closed with a message from Daniel, who began by acknowledging that people were tired—tired of Covid, tired of losing, tired of seeing their friends beaten and killed. People were aching for justice and change. "For those of you who have wondered if there are others out there," he said—and again I found myself weeping—"let this be a moment of refuge. Savor the moment of togetherness. Savor the beauty of each other and our music, our poetry, our dancing, our stories, our tears, our laughter, our grace for each other. It is our love and our humanity that can redeem what we have done to the Earth. There is no limit to the passion and fierceness of love." Acknowledging that there were many others out there committed to the status quo, he called us to recommit to this work. "Our passion for life has to exceed our opponents' passion for greed," Daniel concluded.

SEEING THE number and variety of people doing this work around the world was inspiring, especially given the risk many were taking in their own countries. Their courage and love encouraged me to take greater risk myself by joining a civil disobedience action led by Gina Peltier and other Indigenous women in the aftermath of the Line 3 campaign. They staged a sit-in of the Bureau of Indian Affairs in DC, fifty years after another such Indigenous action. This time they made the connection between colonization and climate destruction, while sitting in a circle with sage burning and volunteer medics checking that diverse participants had what they needed. Arrests were conducted by Homeland Security agents, who broke Gina's thumb and roughly dragged away several other Indigenous leaders. The group faced the most serious potential charges of my activist career, although they were ultimately not filed.[232]

The action itself connected many struggles, including people who had been at Standing Rock and Line 3. Just before the action, I was able to introduce Gina and Sharon Lavigne, who was in DC speaking about St. James Parish at other protests taking place that week. Later at an environmental justice action in Chester, Pennsylvania, a Swarthmore College student approached me to say that she'd been in the cell next door after our DC arrests. We met in Chester because EQAT had begun a new campaign focused on the money that supports polluting industries. It turned out that money connected my home state to every other place I had been on this journey.

CHAPTER TEN

APPLYING THE LESSONS IN THE VANGUARD CAMPAIGN

In April 2022, members of Earth Quaker Action Team prepared to begin a five-day walk from the industry-laden waterfront of the Delaware River to the global headquarters of Vanguard, one of the largest asset managers in the world. Excitement was high as we unloaded red-and-white banners that read, "Vanguard invests in climate destruction." Residents of Chester would be joining us shortly in front of the local incinerator they'd been fighting for thirty years. Together we would make the connection between pollution, climate change, and the companies profiting from this dangerous way of doing business. Our walk corresponded with the launch of a global campaign against Vanguard that included larger organizations like the Sierra Club.

While targeting banks was now an established strategy, going after asset managers like BlackRock and Vanguard was a newer approach to removing money as a pillar supporting fossil fuels. On behalf of their clients, these companies purchase trillions of dollars' worth of corporate stocks and bonds. Although our long-term goal was to get Vanguard to stop investing in fossil fuels, there were many ways they could use their power in the shorter term. Vanguard could vote for climate-friendly shareholder resolutions, and make it easier for its customers to invest in funds that were climate friendly. In the medium-term, Vanguard could encourage companies in its portfolio to make plans to transition away from fossil fuels.

EQAT was asked to anchor the grassroots effort in Vanguard's backyard just as we were wrapping up work on the Power Local Green Jobs campaign. The invitation felt serendipitous as well as strategic. Vanguard was based in suburban Philadelphia, while many financial institutions were based in New York. The fact that our Quaker base included older investors made our demographics a strength in this campaign. We knew that people saving for the future didn't want that future threatened by climate chaos, though many were inadvertently invested in the companies most responsible for it.

GOING AFTER the money behind climate destruction made even more sense to me after seeing firsthand how hard it is to stop one destructive project at a time. Curious about Vanguard's role in places I had visited, I discovered that it was the number one investor in Enbridge, which built the Line 3 pipeline across the Mississippi headwaters, and Exxon, on the northern end of Cancer Alley. Using Yahoo Finance to look them up in turn, I found that Vanguard was one of the top investors in almost every company I heard complaints about in Louisiana, including Formosa, Shell, Entergy, and BP. Vanguard was also the world's top investor in coal companies like Peabody, which had desecrated Black Mesa, and Adani, which was huge in India. I started joking that God knew that I'd be campaigning against Vanguard when way opened for me to go to all these places. Or maybe, Vanguard was just that big. Likely both.[233]

As EQAT began this new focus, our campaign director got another job, and the board asked me to step in as interim director. It immediately felt like the right role at the right time. Challenging the money behind companies causing harm was a way I could show solidarity with the communities that had taught me so much. It was also an opportunity to apply some of what I'd learned from my travels.

OUR VOLUNTEER research team looked up the ownership of polluting industries in our region. We found a concentration thirty miles south of Philly in and around the town of Chester. Along just a few miles of Delaware River waterfront, an oil refinery, a chemical plant, a paper mill, and an incinerator were all owned by Vanguard portfolio companies, or their parent companies. I had visited the predominantly Black community and witnessed how close the trash incinerator was to the redbrick row homes. Now, I learned how much burning trash contributed to both climate change and severe health risks. People in Chester had fought back for decades, winning a groundbreaking campaign to shut down a medical waste facility, along with other limits on industry. This made their community the perfect place to begin a forty-mile walk intended to physically make the connection between Vanguard's profits and the consequences of business as usual.

BUILDING CONNECTIONS

In January, I reached out to Zulene Mayfield, the founder of Chester Residents Concerned for Quality Living (CRCQL, pronounced *circle*), which was campaigning against the trash incinerator owned by Covanta. I explained our campaign against Vanguard and what we had learned about its investments in Chester. I asked if Zulene would be open to having EQAT launch our April walk in her community. She said she was familiar with Vanguard, but wasn't sure if others in Chester would be, since not everyone had a job with retirement benefits. I offered to highlight

CRCQL in our media outreach or do other things to help make the event useful to them. Zulene was open to our coming, but seemed skeptical, which I did not take personally.

. One of the things I'd learned from other frontline leaders was that there were very good reasons for them to be skeptical of outside groups that showed up late to their struggles. Some organizations wanted a photo op more than a relationship. Others worked alongside frontline communities but then got top billing in press accounts of their accomplishments. Another frequent source of resentment was unequal funding. In recent years, as environmental racism has gotten more attention from the press and from granting agencies, predominantly white groups have added environmental justice to their missions, paying for offices and salaries with grants while frontline groups still scramble for funds. Other power disparities can also add to mistrust. I remembered that it took two years for the people of India's Mahan Forest to trust Priya Pillai, a Greenpeace lawyer trying to connect climate change to their local issues.

We were asking Zulene to host an event with us in less than four months, aware that this was short notice. There were several things that helped to make it happen. Most importantly, Zulene knew that her community needed allies, so she was willing to give us a chance. I understood that hosting an EQAT action wasn't the most important thing on her to-do list, especially since they were planning their annual environmental justice walk for the day after ours ended. It also helped that Erica Burman, a key CRCQL member, was a Quaker familiar with EQAT. Erica joined our calls and acted as a liaison when Zulene was busy. Spiritually, I worked on letting go of what I wanted in case CRCQL decided to say no. I knew companies often engaged in "consultation" with communities without any intention of respecting their wishes if they didn't align with what the company wanted. It was important that climate groups didn't replicate this pattern.

After Zulene confirmed that they would host the start of the Vanguard's Big Climate Problem Walk, CRCQL invited EQAT members to support them at a few upcoming events. I realized that a number of Quakers who lived in or near Chester in Delaware County had long shown up to support CRCQL, which also aided our organizational relationship building.

ON A rainy March day, I drove forty-five minutes to a meeting of the board responsible for Delaware County's waste. Packed around the edge of the small room were Chester residents and their supporters, calling for alternatives to extending Covanta's contract. Most of the small board was suffering from the illusion of separation, as if the incinerator pollution didn't affect their own neighborhoods a few miles away. One older white man tried to dismiss criticism by implying that lead poisoning made people in Chester stupid, reminding me of the whites in Norco who employed racist stereotypes to dismiss the legitimate grievances of their Black neighbors. As the board dispassionately discussed the cost of alternatives, I borrowed a pen and drew a makeshift sign that read, "How much is a human life worth?" Others had come prepared with signs, including local college students.[234]

CRCQL's long-standing and closest allies were college students who founded Campus Coalition Concerning Chester (C-4) in 1996 to support CRCQL's work for environmental and racial justice. With members from several area colleges, C-4 consistently showed up to events like the waste board meeting and a county council action the following week, where CRCQL leaders were planning a die-in, a tactic where people lay down on the ground to dramatize the deadly effects of the incinerator. During a Zoom planning meeting before the action, a C-4 leader from Swarthmore said that the students would not lie down outside the council meeting because they were not the most impacted by Covanta's pollution. I understood that the students were trying to be sensitive to the fact that attending an elite college gave them class privilege, even though their group was racially diverse and young, demographics more vulnerable to the climate impacts of the incinerator.

Based on the comment about not lying down, I suspected that our groups used different strategies to navigate race and other identity issues, with C-4 putting more emphasis on disparities and EQAT emphasizing our common stake, even though both groups recognized both. I started to wonder how CRCQL members would feel about EQAT's predominantly white members. I felt reassured when someone said it would be nice to sing at the action, if they could find a good song lead. I mentioned that our member Ingrid Lakey could play this role if they didn't mind a white song lead. Two older Black women from CRCQL unmuted to say that they didn't care about the song lead's race as long as she could sing!

The action went well with good turnout from each of the groups. The students took the riskiest role, sitting in front of the county office doors. Speeches about the cancer and asthma that shortened people's lives in Chester tied directly to the die-in tactics. Near the end, Ingrid led a rousing version of the song "Woke Up This Morning (With My Mind Stayed On Freedom)," inserting "clean air" for freedom in one of the verses.[235] One of the speakers from Chester was moved and said her mother loved that song during her involvement in Chester's civil rights movement.

I was surprised to learn that Chester was known as the "Birmingham of the North" during the freedom movement. In fact,

George Lakey, one of EQAT's founders, was a young man when he was arrested for the first time nearly sixty years earlier during the campaign to desegregate Chester's public schools. He was one of many people who were beaten by the police.[236] I also learned that it was during the 1960s that local college students began showing up as allies to people in Chester, making C-4's partnership with CRCQL part of a longer story of solidarity.

LEARNING ABOUT MY OWN BACKYARD

Four miles south of Chester is Marcus Hook, previous home to a Sun Oil refinery and now a methane gas hub. Before and during World War II, many Black workers moved to Chester for jobs at Sun and its segregated shipyard, some migrating from the South hoping for greater opportunity. They were the first to be laid off after the war, which was when my father got a job on a Sun Oil tanker. Because of job discrimination, white flight, underfunded schools, and many of the same policies that shaped Cancer Alley, Chester became 75 percent Black with a per capita income under $22,000. Marcus Hook, which is over 90 percent white, is only slightly more affluent than Chester. In contrast, Malvern, home to Vanguard's headquarters, has a per capita income four times Chester's with homes five times as valuable.[237] There has been some collaboration between people in Chester and Marcus Hook, but their communities remain more separate than the air above them or the water that flows past both riverfronts. They reminded me of Norco, Louisiana.

The policies and ideas of colonial conquest shaped the Chester waterfront as much as anti-Black racism. One of the demands of the campaign was that Vanguard write an Indigenous rights policy, so it felt especially appropriate to invite a Lenape speaker to the walk launch. For several years, I'd wanted EQAT to build a relationship with local Lenape, also called Lenni-Lenape. A few of us had been pursuing this as individuals, but we hadn't acted as an organization partly because the history of this region made the current dynamics confusing.

BEFORE COLONIZATION, Lenape territory, known as *Lenapehoking*, spanned the lower Hudson Valley down both sides of the Delaware River to where it meets the Atlantic Ocean, through what would become New Jersey, eastern Pennsylvania, and Delaware. In what is now Philadelphia, a large and stable population of Lenape lived along the rivers and creeks, planting crops, hunting, and fishing. Dutch, Swedish, and Finnish settlers brought the first enslaved people to the Delaware Valley. They also carried smallpox, measles, and influenza, strange new diseases for which the Lenape had no natural immunity. Chief Dennis Coker of the Lenape Indian Tribe of Delaware says that his community has documented six smallpox epidemics during the 1600s, which killed 90 percent of his people. "By 1700, our people were just bewildered and shocked," he said, speculating that this tragedy was part of what made them open to Christianizing.[238]

In 1681, England's King Charles II granted the colony of Pennsylvania to William Penn, partly to settle a debt he owed to Penn's deceased father, and partly to give Penn's troublesome Quaker community a reason to emigrate. In a clear example of the Doctrine of Discovery at work, the king specified that Penn and his heirs had the right to all the rivers, islands, inlets, fish, whales, mountains, woods, fields, mines, quarries, gold, silver, stones, metal, and anything else, discovered or undiscovered.[239]

Penn conceived of the Pennsylvania colony as a "holy experiment," free from religious persecution, and the port city of Philadelphia as "a green country town." According to Quaker mythology, he endeavored to treat the Lenape fairly, and many argue that he did better than other European Christians. But Penn didn't question the logic of an English king offering him other people's land. He was a real estate developer, who sold the suburban area where I grew up to a group of Welsh Quakers before he even left England. In 1682, he named the first place he landed in his new colony Chester, after the English home of one of his companions.

When he did make an agreement with the Lenape, Penn assumed that they shared his European view of land ownership. The Lenape believed they were making treaties to share land, where they would continue to live, hunt, and fish. Even before Penn's sons expanded Pennsylvania through outright and malicious deceit, thousands of Quakers took advantage of the epidemics decimating Lenape communities. One wrote that the Lord "was removing the heathen that know Him not, and making room for a better people."[240] This cynical interpretation reminded me not to assume that what I experience as way opening is always from Spirit.

As the settlement of Lenapehoking accelerated in the eighteenth century, most surviving Lenape were forced west. Many were massacred in Ohio. Some then moved to Canada. The largest population of Lenape eventually ended up in Oklahoma. Today, Pennsylvania is one of the few states with no federally recognized tribes, no state-recognized tribes, and no reservations. The federal government recognizes three Lenape tribes, all living outside Lenapehoking, and called "Delaware" after an English governor of Virginia. Small Lenape communities survived in New Jersey and Delaware, eventually receiving recognition from those states. A

Pennsylvania-based group is working to reclaim Lenape heritage after their ancestors intermarried with white settlers, but other Lenape strongly reject their claims, urging groups like EQAT not to partner with them. It is a painful situation, directly rooted in colonization.

After conferring with others more familiar with these dynamics, I decided to reach out to Chief Coker of the Lenape Indian Tribe of Delaware. I'd met him a few times through educational programs, as Quakers increasingly grappled with our role in colonization. I had also joined in a cleanup day for a piece of land his community had been gifted, full of trash and invasive species. When I invited Chief Coker to the walk but didn't hear back, a mutual friend said to assume he was busy leading a community with many needs and few resources. Understanding this was true of many frontline leaders, I trusted that there would be another opportunity to include his community, if they wished.

When we gathered to launch the walk to Vanguard, I took in the damage colonialism had wrought. From our meeting spot, I couldn't even see the Delaware River waterfront, which was flanked with polluting industries. Less than a year earlier, I had sat along the banks of Minnesota's Red Lake River with Gina Peltier, who said that every river and stream in Turtle Island was clean before the colonizers came. Unlike the air, the connection to that history was clear in Chester.

THE WALK

The Vanguard's Big Climate Problem Walk began in the shadow of the Covanta incinerator. Serving as spiritual anchor for EQAT, Dwight Dunston invited us all to take three deep breaths together. Dwight said that breathing mindfully was itself a radical act, something our ancestors didn't necessarily have the opportunity to do—for themselves or with different kinds of people. He reminded us that we were on the ancestral land of the Lenni-Lenape, who lived in right relationship with this land before many were forced to move away. He acknowledged that many Africans were also forced to leave their homes, brought here in enslavement. A Black Quaker artist and teacher of Kingian nonviolence, Dwight said that because history is a continuum, these events shaped the injustices around us today, but we still have hope. "We have still chosen to love," said Dwight, who added that love doesn't mean not holding people accountable. It means doing so with intention, groundedness, and a bigger vision.

As the emcee, I introduced each speaker from Chester and connected what they said to the reason for our five-day walk to Vanguard. Kearni Warren said that her great-grandmother migrated with her grandmother to escape the hardships of the deep South, not realizing the injustices their children would experience in Chester, where both of her parents had cancer at the same time. After Kearni named the many polluting facilities in the region, I pointed toward the Monroe oil refinery (owned by Delta), the paper mill (owned by Kimberly-Clark), the chemical plant (owned by Evonik), and further upriver, the Boeing plant and the Eddystone gas plant. After each one, I asked the crowd to guess who was invested in the company. "Vanguard!" everyone shouted. I explained that Covanta had recently been purchased by a Swedish company, but Vanguard was still invested in the incinerator that burned trash from New York, Philadelphia, and Delaware County.

Zulene Mayfield spoke about the chemicals released by the incinerator, naming both racial disparities and the common stake of everyone in the crowd. "It has degraded our quality of life, our home property values, and totally destroyed any sense of community," she said, pointing out the wooden front porches where Chester residents chatted with their neighbors when she was growing up. "We have residents who have been told by their doctors not to even open up a window," she said, challenging industry's indifference to the lives of people who don't look like them. "We have the frickin' right to breathe!" To those who thought they had escaped this pollution, she said, "We may get it first, but you breathe the exact same crap we breathe. We all have a dog in this fight. Ain't no force field over Chester that keeps all of the nasty air in here." Both EQAT and CRCQL members shouted, "That's right."

"Nothing man-made will stand forever," Zulene declared. The incinerator towers would fall, "And we will help facilitate it." She closed by acknowledging that when I first called, she didn't know who EQAT was. "But I know now. We know that there are other like-minded people that give a damn about what happens to people." She said that this kind of solidarity and bridge building was scary to power holders.

To build on Zulene's point about the nasty air spreading, I shared that my daughter had her first asthma attack at five years old. When I asked the emergency room nurse what caused asthma, she said it was probably pollution. We lived in a Philadelphia neighborhood with many large trees. I didn't know we were breathing polluted air, but people in Chester knew. "It's not equal, but it's shared," I said, noting that a Covanta spokesperson had told *The Philadelphia Inquirer* that we were outsiders.[241] "Does anyone ever call Covanta outsiders? Or Kimberly-Clark?" I asked. The crowd laughed.

Speaker Will Jones, also of CRCQL, noted that EQAT being called "outsiders" was a classic divide and conquer tactic. "They are trying to isolate us and keep us contained and separated without allies." To people who were just learning about these issues he said, "Once you know, you can't put your head under the covers." As we prepared to start walking, George Lakey told the crowd that walking was an old activist strategy, a way of "gathering energy." It builds our courage and trust in ourselves and each other. We set off with our red banners, spirits high.

ON THE second morning of the walk, we visited Eastwick, a Philadelphia neighborhood, where increasing flooding from climate change was bringing toxic waste from a nearby landfill into people's brick homes. Due to relentless advocacy by the community, the EPA was finally cleaning up the toxins.[242] By the following afternoon, we reached the suburbs where we celebrated Passover with longtime activist Rabbi Arthur Waskow, who blames the plagues of climate change on "climate pharaohs," like Chase Bank and Vanguard. A hundred people gathered outside a Chase branch with wine, matzah, song, and dance, pointing out that Vanguard was the top investor in Chase. The next morning, we stopped at the stately home of Tim Buckley, then CEO of Vanguard, where two of us delivered a bouquet of flowers and a card, explaining that we were praying for him to use his disproportionate power for the common good. His shades remained drawn and the door unanswered.

On the final morning, Earth Day, we arrived at Vanguard's global headquarters with close to 150 people, including friends from POWER, our partners in the Power Local Green Jobs campaign. Bishop Royster spoke about the need to divest from dirty energy and "invest to build a world that truly works for our communities." A new grandfather, he declared, "We need all hands in to transform this planet! Whether you're Black, you're white, you're Brown, you're Indigenous, you're Asian Pacific Islander;

you're urban, suburban, rural; Democrat, Republican, socialist, Working Families; whether you run a corporation or you are an underpaid worker, whoever you are, this Earth is your home! We all belong. *We need each other.*" I remembered saying these last words in New Orleans East when climate advocates from around the world met over pho.

Song verses like, "We are young and old together," and "We rise," helped to build a sense of connection within the crowd. Speakers representing youth and elders delivered a Sierra Club petition with several thousand signatures to Vanguard's driveway. Most participants stood on the grass, while those willing to commit civil disobedience spread out across the Vanguard driveway. Over a dozen people took turns reading statements before they laid down in the crosswalk. I helped to write the statements, each of which acknowledged a community hurt by and/or resisting companies in Vanguard's portfolio. Some were in places I had visited, like Louisiana and India. Some were not, like Uganda and Tanzania, where people were trying to stop a nine-hundred-mile crude oil pipeline that would displace fourteen thousand families and many unique species.

The police stood by. Clearly Vanguard did not want the added publicity that would come with arrests, so they diverted traffic to their other three driveways. Still, we knew we had gotten their attention. That week, Vanguard put out its first-ever statement acknowledging, "Climate change, and the ongoing global response, will have far-reaching economic consequences for companies, financial markets, and investors."[243] Vanguard also announced a new climate-friendly fund. These small steps showed that they cared about their public image, though with such a huge company, we would need to build more power to motivate bigger change.

BUILDING POWER WITH LOVE

Over the following months, the global Vanguard campaign organized customer call-in days and challenged Vanguard's branding in
the press and at public events. Member organizations also worked
to organize employees at large corporations, like Amazon, that
used Vanguard to manage their retirement funds. Since Vanguard
refused to meet with any of our coalition partners, seven people
from EQAT and XR Philly walked onto Vanguard's campus and
asked for a meeting to discuss Vanguard's plans for climate steward-
ship. Instead of a meeting, the seven got arrested, some Vanguard
customers. We used a series of smallish actions to train more people
to play roles, such as action lead, police liaison, and song leader. The
next time we engaged in civil disobedience, we had enough leaders
to block all four Vanguard driveways, emphasizing our need to
interrupt business as usual. This time sixteen people were arrested.

Founded in solidarity with Indigenous people protecting the
Amazon Rainforest, the group Amazon Watch organized a delegation from Peru to visit US corporations financing oil extraction.
EQAT and allies greeted them at Vanguard headquarters with
more than 120 supporters. "Are these people all here for us?"
asked Nelton Yankur through a translator. Nayap Santiago told
the crowd how Petroperú, a state oil company, was hurting their
people and land, with oil spills contaminating an area recognized
by the United Nations for its biological diversity. Now, Petroperú
was raising money for oil expansion by selling bonds, with help
from Vanguard. "They do not realize that through their investments, they are affecting us very directly," Nayap said of Vanguard,
which had ignored their request to meet. He mentioned the global
climate negotiations that were taking place in Egypt, and how
they had not helped their Indigenous community.

As emcee, I said that if the world's leaders were negotiating deals that didn't stop the expansion of oil in the Amazon, they were not fixing the problem. Other speeches reflected themes I'd heard across the United States: the need for Indigenous people to have sovereignty over their territories; the way industry sows conflict within communities; and the need for unity. The delegates were modeling this, bringing together the Achuar Federation and the Wampís Nation, as well as the coastal fishermen, whose livelihoods were also impacted by oil. Those of us from Pennsylvania were repeatedly addressed as *"hermanos y hermanas,"* or "brothers and sisters."

Before their arrival, I heard that the delegates would appreciate meeting with local Indigenous leaders. Chief Coker agreed to drive up from Delaware in pouring rain to meet them. The director of Pendle Hill retreat center, Francisco Burgos, offered to host in his residence, and I showed up with salmon and other cooked food. Amazon Watch staff translated as the Peruvians described their home, including the plants that were important to their spirituality. Chief Coker described Lenape history. I felt honored to witness this connection being made, but also felt the challenge of maintaining such relationships across distance and language.

TO DEEPEN our own commitment, and to build pressure on Vanguard's CEO, we decided to hold Quaker-style silent worship outside his suburban home during a global week of climate action organized by GreenFaith, which brings together people of diverse faiths. In conversation with a Quaker who was uncomfortable confronting someone at their home, I mentioned that eighteenth-century Quaker abolitionist John Woolman often felt led to visit Quaker enslavers at their homes and challenge them in a direct but loving way. After mentioning Woolman, I did an internet search on him and discovered that the 250th anniversary of his death was the week of our action, so we planned it for that day.

Like EQAT, Woolman was distraught by a great wrong and sought to be faithful to divine guidance, even when it led him to do things that were hard or scary. For him, this included giving up cotton, dyed cloth, and silver platters, since they were produced with enslaved labor. A knowledgeable Quaker friend said that Woolman told the enslavers he visited that he genuinely worried that living in leisure off the labor of others was bad for their souls. After his visits, many freed the people they enslaved. While moved by this example, I was also acutely aware that our interconnected economy made it impossible to eliminate what Woolman called "the seeds of war" in our possessions.[244] That was part of the reason why EQAT was trying to change the policies that supported the industrial growth society rather than focusing on individual change.

Early in my time in EQAT, George Lakey told a story that helped me understand the difference between changing an individual and changing a system. He said that during the Vietnam War, activists thought that if they could just convince Secretary of Defense Robert McNamara that the war was wrong, he would end it. Turns out, McNamara knew the war was wrong—and unwinnable. When that knowledge became too uncomfortable, he didn't end the war. He quit his job. We'd heard that Buckley was concerned about climate, but we also knew the institution was bigger than one person. So we were pressuring both. I kept coming back to Dr. King's assertion that we need love and power to make real change. To me, that means many, many groups pushing on the pillars that support the status quo, while inviting those with more power into our vision of a just and sustainable alternative.

EQAT was experimenting with how to do this. On the day of the worship action, fifty people sang as we carried folding chairs from the train station to Tim Buckley's quiet street. After a brief introduction about John Woolman, we sat in silent prayer across from Buckley's large home, where the shades were again drawn. British Quakers brought even more people to Vanguard's London headquarters, while Quakers from across the United States prayed on Zoom, which connected the three synchronized actions.[245] Many of the 150 total participants reported that the worship time felt deeply grounding. I found myself praying for Tim Buckley to have courage, as well as the rest of us.

Even if he had to hear about it from his security staff or his wife, it felt like coming to Buckley's home was a way to puncture his illusion of separation from the suffering caused by corporations that would take a meeting with him, though not with us. The best-case scenario for this campaign was that Vanguard would use its tremendous power for the common good, which of course included its customers. Engaging customers was yet another way to build power with love.

THE GREENWASHED PATH

Early in the campaign, Vanguard customers started asking us if there were Environment, Social, and Governance (ESG) funds they could choose that would be better for the planet. We wanted Vanguard to help change companies like Exxon and Enbridge, not just exclude them from a few small funds. Even if that were our strategy, there were real issues with ESG ratings, which theoretically help investors identify companies that are doing well by doing good. For example, reducing packaging or energy use are both good for the Earth and save the company money, especially in the long run. Unfortunately, when it comes to environmental and climate concerns, many companies "greenwash" their practices, making them sound better than they actually are.

Billed as a climate change solution, there is a plan to sequester carbon emissions underground along the Louisiana coast, which is prone to flooding. Locals fear the gasses might escape through abandoned drilling rigs. Unsurprisingly, carbon capture is favored by the fossil fuel industry.[246] Exxon has gotten ESG credit for pursuing this technology. When Covanta was purchased by a Swedish company, they claimed the incinerator business would be good for ESG scores, since it is a non-fossil fuel energy source that keeps trash out of landfills. Even assuming the incinerator has reduced air pollution since the 1980s, as claimed, burning trash still creates emissions that are bad for local people and for the climate.[247] In a classic case of greenwashing, its 2024 sustainability report rebrands the company "Reworld" and features a photo of a Black woman's hand reaching out to a blue butterfly.[248]

At a conference on climate finance, I met two Chilean activists, who were fighting a greenwashing scheme by Arauco, another company in Vanguard's portfolio. A violent coup in the 1970s enabled the privatization of many Chilean resources, including the land of the Mapuche people. Some of the world's oldest trees were felled to make room for monoculture timber plantations that use a dangerous amount of water. Over fifty years, Arauco's industrial plantations continued to expand, further impoverishing the region's biodiversity, and drying the forest just as warming is increasing forest fires. The Chileans explained that the company is now getting subsidies to make wood pellets and selling them as biomass, which counts as renewable energy on their ESG score. Their story brought home to me why asset managers like

Vanguard need Indigenous rights policies, which would make it harder for such schemes to be passed off as green energy.[249]

THE GROUP As You Sow tries to help investors understand where their money is really going. I used their Fossil Free Funds tool to look up Vanguard's ESG offerings and found that most got dismal report cards, especially if you wanted to avoid deforestation as well as fossil fuels. The only Vanguard fund that got an A for an absence of fossil fuels was the Vanguard Baillie Gifford Global Positive Impact Stock Fund, the one launched in response to our campaign.[250] I looked up my own retirement plan through a different asset manager. I'd assumed the "social choice low-carbon fund" was good, but when I read the fine print, I found a small percentage was invested in ConocoPhillips, one of the companies responsible for the high levels of dioxin in Mossville, Louisiana. Following the money was puncturing yet another layer of my illusion of separation.

Given the climate movement's criticisms of greenwashing, I was caught off guard when conservative politicians launched a backlash against ESG, disparaging it as "woke capitalism." As one financial commentator explained, "By labeling ESG 'woke,' conservatives imply that large parts of the US $100 trillion global asset management industry have been hijacked by leftists. Having spent time with lots of asset managers, it's nonsense."[251] Still, the "woke" trope gained traction, tapping established narratives about racial equity and climate change as left-wing plots.

New Hampshire Republicans tried to make it a felony for state funds to be invested using ESG criteria. Some Republican-led states started to move their money out of BlackRock, Vanguard's biggest competitor, which had done more to encourage ESG investing. Meanwhile, New York State was moving some of its billions away from fossil fuels and toward low-carbon investments on the grounds that this was smart financially as well as better for the climate. In a political twist, Republicans were suddenly the ones accused of being anti–free market.

I took the "woke capitalism" debate as a sign that the climate movement's strategy of pressuring the financial pillar of power made elites worried. Indeed, some analysts said that the backlash was prompted by a successful attempt to elect two proponents of energy transition onto Exxon's board. BlackRock and Vanguard had supported this effort in 2021, demonstrating the power that asset managers have.[252] Now, faced with backlash, Vanguard retreated from its only tepid climate commitment in December 2022. We had intentionally given the asset manager a chance to take more positive steps before encouraging customers to move their money. In early 2023, we decided it was time.

STANDING AGAINST CORPORATE GREED

We always knew that if we were going to divest from Vanguard, we wanted to do it together, as a form of collective pressure, so we asked customers in our network to phone and write the asset manager first. As we escalated to moving money, we offered webinars to help people prepare. We couldn't pretend to be financial advisers, but we supported investors to identify their priorities and find alternatives, using the As You Sow tool. We soon realized that moving money was intimidating for those who had been taught that the market was beyond their understanding. For some, the process brought up fears of not having enough, especially given our country's fractured safety net. For others, money moving was a good impetus to consider reparations, the practice of repairing past harms with compensation to those who have been hurt, or their descendants.

As an organization, our strategy focused on moving money out of Vanguard and its funds as a way of building collective power, but we soon realized that even talking about money and encouraging people to make more deliberate choices was countercultural, especially within a group with a wide economic and generational range.

GEORGE LAKEY often points out that extreme economic inequality is at the root of our political polarization. He also asserts that times of polarization slacken the status quo, like a blacksmith's fire, enabling change, which can bend in either direction depending on the power of social movements.[253] When I let go of the Democrat/Republican binary that has rigidified our political thinking and thought about all the people who wanted a world that wasn't shaped by corporate greed, it expanded the number of potential allies I could imagine. Certainly short-term, profit-centered thinking has hurt American workers, from driving jobs overseas to cutting worker benefits. The rollback of pollution limits also threatens industry workers. While organized labor was an important part of the coalition that passed climate legislation in New York, EQAT's outreach to unions during the Power Local Green Jobs campaign had been unsuccessful. This was a potential growth edge for much of our movement.

So, when I heard that the United Mine Workers of America (UMWA) were visiting Vanguard headquarters in June 2023, I skipped a staff meeting to listen to their perspective. I entered the small park where EQAT often gathers to find five hundred people who had traveled from as far as Alabama, West Virginia, and Ohio. Many were coal miners, retirees, their families, and allies. I learned that when coal prices were low a decade earlier, an Alabama coal company pulled itself out of bankruptcy by soliciting capital invest-ments and steep pay cuts from the union. When coal prices and profits rose, the money flowed to investors and company execu-tives, not to the workers, who had been promised a pay raise. The renamed Warrior Met Coal not only reneged on its labor agree-ment, it also hired nonunion workers during and after a strike, and tried to get the union decertified. After years of struggle, the union had come to Pennsylvania to ask Vanguard—one of Warrior Met's largest institutional investors—to use its power to help workers get a fair deal.[254]

I was fascinated by the similarities and differences between our groups. The mine workers filled the parking lot with large pickup trucks rather than EQAT's usual row of Priuses. They had a US flag on the stage. The one time we brought a stage to Cedar Hollow Park, we borrowed it from a queer cabaret troupe called the Bearded Ladies, so it was decked in silver, pink, and purple, instead of red, white, and blue. Yet, when I listened to the union speeches, I felt there was much we agreed on. Longtime President Cecil Roberts told his members that they could not discriminate against anybody because of race, color, national origin, language, religion, or gender. This was the proud legacy of the UMWA, which refused to be divided when coal companies brought in Black and immigrant workers to break the union in its early years. Looking out on the enthusiastic crowd, I saw many signs that said, "We Are One."

A veteran of the Vietnam War, Roberts asked all veterans to stand. Many in the crowd did. Then he asked anyone who had family who were veterans to stand. Nearly everyone stood, including me, thinking of my father and cousins. "Now I want Vanguard to take a look at this. We've been doing all the fighting and all the dying so capital can survive," he said. "We're sick and tired of you taking all of the riches from this country." Noting that Vanguard leaders had refused to meet with them, he described them as "people who never went down a two-thousand-foot shaft. They go to the fanciest restaurants to eat. Our guys are opening their dinner buckets three miles underground. The gap between us and them needs to be closed," Roberts declared, saying that no company should control trillions of dollars.

Marching down to Vanguard's driveway with the union, I thought of all we had in common. We all wanted to see human lives valued over profit. We also shared a willingness to act collectively, even to risk arrest. One speaker said, "We must redeem the world," which sounded like an EQAT line. But there were also real obstacles to finding common ground between us, and not just because of the vehicles we drive or the range of pronouns we use. The mine workers want to hold on to coal jobs, while we feel the urgent need to move off of coal and other fossil fuels. We all care about our children, but I'm worried about the next superstorm to hit New York, where my children live. One of the retired miners I spoke to shared that his fear for his children was fentanyl, which had ravaged his economically depressed West Virginia community. EQAT had started campaigning for green jobs in 2015, and several years later they were booming nationally. But they had not yet replaced the income lost in places like West Virginia, or the Navajo Nation, for that matter.

Like Vanguard executives, I've never been down a two-thousand-foot shaft, though my great-grandfathers knew that life. Thinking about the span of generations reminded me of the Haudenosaunee teaching that we have to think, not just about our own children, but about seven generations into the future. From that vantage point, what our descendants would have in common seemed much more important than what kept us apart today. It was also a perspective that showed the shortsightedness of using quarterly profits to assess value.

BROADENING AND DEEPENING

We launched our money moving effort in Chester in front of the incinerator. Between speeches and songs, members of CRCQL, C-4, and EQAT wrote postcards to Vanguard's CEO urging him to change course. The postcard featured a map of Vanguard's investments along this stretch of the Delaware River, including a new liquefied natural gas facility, still in the preliminary stages, but with several Vanguard portfolio companies vying for roles in the project. Such hazardous facilities had already been built along the Gulf Coast. Soon after the Chester action, we brought our messages to Vanguard headquarters. Employees leaving the annual company picnic rolled slowly past our banner announcing an initial $17 million moved, as speakers explained why they were leaving Vanguard, some after decades as loyal customers.

We tracked the growing amount of money moved with a graphic on our website that showed our total was just "the tip of the iceberg." As more customers joined the effort, most found it was a low-risk way to chip away at the financial pillar supporting climate destruction. The Never Vanguard pledge on our website also included options for those who were not customers, such as students who pledged to never work for the company, as long as it stayed on its destructive path. Local college students started organizing to challenge Vanguard when the company tried to recruit on their campuses. EQAT started broadening the campaign in other ways, too.

In the summer of 2023, a few of us traveled to Oregon for a national Quaker gathering organized by Friends General Conference (FGC). We offered a workshop on nonviolent direct action and planned an action with the help of local Quaker climate activist, Cherice Bock, whom I'd met through Line 3 solidarity. We found that Vanguard was invested in the Pacific Northwest's largest deforestation company, as well as another Covanta/Reworld incinerator, this time near a Latino community. There was a local campaign to prevent the GTN Xpress pipeline, which was proposed to run through Idaho, Washington, Oregon, and California—again a Vanguard investment. Plus, Oregon used Vanguard to manage its college savings plan, so the state was impacted by Vanguard's choices in multiple ways. The Salem branch of 350.org showed up to support our action, in which young Quakers led worship. Cherice invited nationally known Indigenous song keeper, Quiltman Sahme, who played the drum and sang a moving prayer.

AFTER THE Quaker gathering, Cherice and I drove with a picnic lunch to visit Quiltman on the Warm Springs Reservation, which was governed by a confederation of the Wasco, Warm Springs, and Paiute tribes. The reservation included a section of the stunning Cascade Mountains as well as grasslands and a tributary of the Columbia River. While the history and geography were unique, I heard in Quiltman's stories many of the same themes I'd heard in other Indigenous communities.

Sitting in a metal building with electricity from off-grid solar panels, Quiltman talked about his mother, who was part of several environmental campaigns, including one to stop logging on the sacred sites of Mount Hood. This sprung from the reverence for Mother Earth that was part of his people's ceremonies and daily practice. I observed that despite their brutal efforts, the US government had failed to kill Indigenous culture. He nodded. His grandfather had hidden one aunt so she could keep the traditional knowledge that was beaten out of her siblings at the boarding schools. He was sent to such a school, and decades later founded a community-based survival school to make sure his own children got a different kind of education. Punctuating his stories with laughter, he reminisced about his decades singing and drumming traditional songs, including with artists such as Jackson Browne and Willie Nelson. Now, he planned to turn his space into a community coffee house.

I asked how this land had changed over his long life. He said that earlier generations burned the berry bushes after harvest, which reduced the risk of the extreme wildfires they experienced today. Earlier generations also lived long, healthy lives with salmon as a key food source celebrated at festivals. I'd read that several Indigenous groups in the Pacific Northwest were focused on protecting salmon. Quiltman said that many dams kept salmon from their spawning grounds, and a leak from a nuclear site near the Columbia River had contaminated the fish.[255] Now many people had cancer, which they never had in the old days.

Diagnosed with lymphoma, Quiltman needed to come home three times for chemotherapy during his months at Standing Rock. A former AIM member, he heard the call to come to North Dakota, so he and some friends sang at a wedding to earn gas money for the journey. He described the police violence at Standing Rock and the disrespect shown to sacred burial grounds and prayer pipes. He also shared how amazing it was to see so many people come together, including many clergy and thousands of veterans. I mentioned a ceremony I'd heard about, where veterans kneeled and apologized to the elders for their role in US government violence against Indigenous people. "It was powerful. Nobody expected that," he said.

I asked Quiltman why he'd agreed to drive a few hours to join our Quaker-led action against Vanguard. "Because she asked me," he said, pointing to Cherice. She had been working for an interfaith climate group two years earlier when they traveled together to the Mississippi headwaters for a Stop Line 3 campaign event. The story confirmed that relationships matter, especially when organizing across difference. One of the ways the crisis of the Earth was bringing people together was through campaigns like Standing Rock and Line 3, often building connections that lasted longer than the specific struggle. I mentioned the prophecies about people uniting during the time when the rivers and fish were poisoned, and Quiltman recalled that one prophecy talked about the "rainbow nation."

I was grateful that my trip to the Northwest allowed me to see some Line 3 water protectors. Earlier in the year, Gigi Nathan had invited me to attend a Sundance, which happened to be in Oregon just after the Quaker conference. I felt honored to be invited and couldn't pass up such serendipity, especially knowing the annual ceremony was an important part of Gigi's healing journey. I headed to the remote location where I spent the weekend praying and being prayed for by a diverse group of people in a ceremony that the US government had outlawed for many years.[256] I remembered what Gigi had told me about healing needing to happen in all directions so people would stop doing harm. With tears streaming down my face, I prayed for Spirit to heal my own illusion of separation—as well as for the Louisiana oil and gas leaders from whom I had felt so estranged during their New Orleans conference.

LESSONS FROM A CONTINUING PATH

When I left EQAT staff to write, I stayed involved as a volunteer as the team prepared for the 2024 national Quaker gathering, organized by FGC. It would be held in the Philadelphia suburbs at Haverford College—the place where I first heard the message about the illusion of separation. It felt like way opening, enabling EQAT to bring Quakers from across North America to Vanguard's global headquarters for an action. Ten years earlier, we won our PNC campaign after a similar confluence mobilized national Quakers, who went home and organized actions in thirteen states. We hoped to replicate that strategy, though Vanguard only had offices in a few cities, unlike PNC, which had branches in many. We hoped to protest Vanguard's investments at other kinds of sites, such as the places where Vanguard portfolio companies were causing harm. While more challenging, this strategy felt like a way to multiply our potential allies while removing the illusion that financial companies were separate from the industries they enabled.

Lina Blount, who had taken over my role as director of strategy and partnerships, built connections with people resisting the Mountain Valley Pipeline, which was built through West Virginia and Virginia and threatened to extend into North Carolina. I joined campaign members in spring 2024 for an Indigenous-led tribunal on the rights of nature and was moved by testimony about the interconnection between river species, such as fish and mussels, which need each other to thrive. I also learned how asserting the rights of rivers and the species that lived in them was a growing strategy to protect all life. The Marañón River in Peru, which has been contaminated by Petroperú oil spills, was found by a court to be a subject with rights.[257] In Minnesota,

Gina Peltier was working on a similar case to honor the rights of wild rice. I thought of the places I had visited and wondered how they might be transformed if the Hudson, Mississippi, Colorado, Columbia, and Ganges Rivers were all protected as living entities with legal rights.

IN JULY 2024, during the FGC Gathering, three hundred people gathered at Cedar Hollow Park as turkey vultures circled overhead. I remembered the animals at my last Line 3 action and Robin Wall Kimmerer's assertion that the humans lining the green path are not alone. We were also joined by water protectors from the campaign to stop the Mountain Valley Pipeline, a bus from the national Stop the Money Pipeline coalition, and Quakers from as far away as Alaska. Lenape elder Walter Durham opened the gathering, which included singing, led by Rev. Rhetta Morgan, and several speeches. A representative of the campaign against the proposed East African Crude Oil Pipeline shared that it had been slowed down in part because African campaigners had gotten banks and insurers to pledge not to support it, with help from global allies, especially in France.[258] Zulene Mayfield spoke about Chester, and Ingrid Lakey tied all of these struggles back to Vanguard.

Daniel Hunter led a brief training, acknowledging the mix of people and styles of activism present. We sang as we walked up to Vanguard's driveway where we held worship that many people found particularly deep. A New England Quaker had made a beautiful clay jar to collect soil from the different places our speakers represented. A week after the large action, a few of our members tried to deliver the jar as a welcome gift to Vanguard's new CEO, Salim Ramji, along with a card signed by nine hundred people. The guards refused to accept the beautiful gift, made with love. We promised to continue showing up for as long as we felt Spirit telling us to.

FOUR MONTHS later, Donald Trump was reelected president. In an unconstitutional power grab, he set about trying to dismantle the US government, especially the parts of it that limit the reckless and abusive power of the fossil fuel industry. By withdrawing the US from global climate agreements, he undermined the systems that were encouraging global collective action. He also exposed the toxic role racism and sexism have long played in our politics, as well as the demonization of immigrants and trans people.

Despite all this, support for climate action has steadily grown, included in states that voted Republican. My visits to frontline communities showed me all the reasons why popular opinion did not always shape policy, but they also strengthened my belief that people can act courageously when they realize that their own lives or their children's are threatened. Inadvertently, Trump motivated many people to take bolder action for the world they wanted to see: a world grounded in love.

I had set out to understand what could build a broader and stronger movement for change. I heard frontline leaders acknowledge that we all breathe the same air and drink the same water, whether we live next to a polluting industry or are invested in one. I came away even more convinced that our problems are connected, and we are more powerful when we are connected, too. I had seen how bravery inspires others, and how people are supported to act bravely through community, singing, and spiritual practices, which can also support us to learn and try again when we fall short. I'd felt my own faith strengthened by the faith and prayer of others, whether those prayers took the form of silence, speaking, drumming, or dance. I also learned from history and the Earth herself that a long-term perspective is key to navigating the backlash that often follows progress. My piece of the green path is small, and I cannot see the whole route from here.

I often remember something Daniel Hunter said near the end of our conversation about navigating race. I asked whether he felt hopeful about the state of the climate movement. In response, he offered a story. Just the day before he was canoeing on the Delaware River with his then two-year-old daughter, and they capsized. "I grabbed my daughter. She was all right, so we began swimming to shore," he recalled. Although they were in a harbor, the waters were very choppy. "There was a moment when I worried that we wouldn't get to shore because we just weren't getting there fast enough. Fear was setting in." I felt my own chest tighten as I listened. This was how many of us felt about our journey toward a life-sustaining civilization. We just weren't getting there fast enough, and our children's lives were at stake.

Daniel said he realized that worrying about the outcome wasn't helpful and wouldn't change his actions anyway. I heard it as a moment of letting go of fear. "There's only one thing to do. I am going to swim to shore," he told himself. "That was it," he said with a laugh, as I wiped my eyes. "So, that's how I feel in terms of where we are." He added that one of the things he's learned from climate colleagues in the Global South is that they are not burdened by the mythology common in US movies and novels, where an individual hero saves the entire universe against all odds. He said it has helped him psychologically to realize that he can't carry the weight of the whole world on his shoulders. That's unfair and unrealistic. All any of us can do is keep swimming, doing our part to reach the shore.

REFLECTION QUESTIONS
AND PLAYLIST

▶ What signs of ecological or climate crisis are you seeing in your own region? Was there any natural disaster, local or not, that made you more aware of these issues?

▶ What have you been taught about your own ability to make change?

▶ Take any power holder you would like to change and draw the pillars that support them. Are there any where you have a special connection, perhaps as a customer, employee, or neighbor? Are there groups already challenging these institutions?

▶ How were you taught to think about nature? Do you consider yourself part of nature or separate from it/her?

▶ Do you think of other species as relatives? What about law enforcement? How does your attitude affect your relationship to the world around you?

▶ Do you know what peoples are indigenous to the land where you live? What more would you like to know about their story, or share about your story if you are Indigenous?

▸ Which aspect of the Great Turning do you feel most drawn to? Stopping the harm, building less destructive systems, or facilitating a shift in consciousness?

▸ What stories from this book did you find most challenging to read? What were most encouraging?

▸ What approaches to navigating race have you experienced? What have you seen help people to find common ground? What have you seen backfire?

▸ If the language of spirituality in this book does not resonate with you, is there another language about our interconnection that does? If you do use spiritual language, how would you describe what ultimately connects us?

For a playlist of songs that reflect the themes of this book, please visit eileenflanagan.com/common-ground/playlist/.

ACKNOWLEDGMENTS

So many people made this book possible. First and foremost are the frontline leaders whose stories are the heart of this book. For some, sharing their experience meant retelling traumatic events. I am deeply grateful for their time, wisdom, and generosity of spirit. Many activists and experts gave background information, insights, and context, which helped to shape my understanding, along with speakers at several conferences. These included the HBCU Climate Conference organized by the Deep South Center for Environmental Justice, the Religious Origins of White Supremacy conference organized by Syracuse University and the Onondaga Nation, the Quakers, Colonization, and Decolonization conference organized by Friends Association for Higher Education, the Yesah Tribunal organized by Seven Directions of Service, and the Global Just Recovery Gathering organized by 350.org and friends.

Thanks to Kumi Naidoo for his kind forward and for all he has taught me about courageous activism over the years. My understanding of nonviolent direct action was forged through my work with Earth Quaker Action Team, which continues to help me learn to practice spiritually grounded and strategic activism. Many EQA-Ters are part of the stories in this book, and I am forever grateful to them. Special gratitude to Dwight Dunston, Daniel Hunter, George Lakey, Ingrid Lakey, and Rhetta Morgan for all the ways you continue to support and challenge me to show up in the world. My Quaker congregation, Chestnut Hill Friends Meeting, provided ongoing spiritual support through this project, especially the accountability group that meets with me regularly: Amey Hutchins, Hedie Kelly, Janaki Spickard-Keeler, and Dylan Steinberg.

This project took shape over several years, and many people helped along the way. Philadelphia Yearly Meeting's granting group provided financial support for much of the travel behind these stories. Further financial support came from the Balthrop Cassidy Family Fund. I enjoyed home hospitality in Mumbai from Nandini Deo and Tim Loftus, with many meals and insightful conversation provided by Neelam and Pramod Deo. Thanks to Rama Subramanian for organizing my stay in Auroville, and to Cindy Yurth and Eric Swanson for hosting me on the Navajo Nation. Cathy Lemann welcomed me on repeated visits to New Orleans, as did Denise and Barney Cassidy in Seattle, Kathy Moser and David Ashton in Albany, Maureen and Mark Murphy in Syracuse, and Anne and Richard Ford in North Carolina.

Thanks to Angelle Bradford for research help, as well as her knowledge of Louisiana. As I tried to figure out how to shape this material, Rebecca Heider was extremely helpful as a repeat reader, as was Nancy Rawlinson. In addition to the people already named above, I am grateful for the feedback, insights, and moral support of Matthew Armstead, Sara Carder, Christine Eberle, Eve Gutman, Greg Holt, Wendy Horowitz, Juan Jewell, Carolyn McCoy, Maureen Murphy, Paula Palmer, Janaki Spickard-Keeler, Betsy Torg, Ellen Tynan, Deb Valentine, Tom Volkert, Sarah Willie-LeBreton, Sara Jolina Wollcott, and Aminata Desert Rose Plant Walker Fire Woman. Any errors or omissions are, of course, my own. Thanks to Striff Striffolino, who designed the images.

The women of Wordspace, and the High Point crew provided cheerleading when I needed it.

Huge thanks to the whole dedicated team at Seven Stories Press, whose commitment to independent publishing is especially important when free speech is under attack. Special gratitude to Claire Kelley, who saw the potential of this book, acquired and advocated for it, and stewarded it through every stage of the publishing process. Further thanks to publisher Dan Simon, whose feedback made it a better book with clearer relevance for our times.

Last but certainly not least, my husband, Tom Volkert, provided many forms of support throughout this long project. Our children, Megan and Luke, continue to provide motivation to not give up on the world. So does Spirit, whose guidance I have tried to follow in this journey, with gratitude for all the ways I was held.

RECOMMENDED READING

Clatterbuck, Mark. *Sacred Resistance: Eco-Activism and the Rise of New Spiritual Communities* (Orbis Books, 2025).

Dunbar-Ortiz, Roxanne. *An Indigenous Peoples' History of the United States* (Beacon Press, 2014).

Flanagan, Eileen. *Renewable: One Woman's Search for Simplicity, Faithfulness, and Hope* (She Writes Press, 2015).

Haga, Kazu. *Healing Resistance: A Radically Different Response to Harm* (Parallax Press, 2020).

Kendi, Ibram X. *Stamped from the Beginning: The Definitive History of Racist Ideas in America* (Bold Type Books, 2017).

Kimmerer, Robin Wall. *Braiding Sweetgrass: Indigenous Wisdom, Scientific Knowledge, and the Teachings of Plants* (Milkweed Editions, 2015).

Lakey, George. *How We Win: A Guide to Nonviolent Direct Action Campaigning* (Melville House, 2018).

King, Martin Luther, Jr. *Where Do We Go from Here?* (Beacon Press, 2010).

McGhee, Heather. *The Sum of Us: What Racism Costs Everyone and How We Can Prosper Together* (One World, 2021).

Mitchell, Sherri. *Sacred Instructions: Indigenous Wisdom for Living Spirit-Based Change* (North Atlantic Books, 2018).

Penniman, Leah. *Farming While Black: Soul Fire Farm's Practical Guide to Liberation on the Land* (Chelsea Green, 2018).

Solnit, Rebecca. *A Paradise Built in Hell: The Extraordinary Communities That Arise in Disaster* (Penguin Books, 2010).

Taylor, Dorceta E. *Toxic Communities: Environmental Racism, Industrial Pollution, and Residential Mobility* (New York University Press, 2014).

NOTES

INTRODUCTION

1. "Jacqueline Thomas Saik'uz First Nation—Forward on Climate Rally Washington D.C. Feb 17 2013," posted February 26, 2013, by Mark Brooks, YouTube, 11 min., 53 sec., https://www.youtube.com/watch?v=4G-N3WB0FP4.
2. Lisa Friedman, "Oil Interests Gave More Than $75 Million to Trump PACs, New Analysis Shows," *New York Times*, November 1, 2024, https://www.nytimes.com/2024/11/01/climate/oil-gas-donations-trump.html.
3. Vann R. Newkirk II, "Trump's EPA Concludes Environmental Racism Is Real," *Atlantic*, February 28, 2018, https://www.theatlantic.com/politics/archive/2018/02/the-trump-administration-finds-that-environmental-racism-is-real/554315/.
4. Christopher Flavelle, "Trump Team Plans Deep Cuts at Office That Funds Recovery from Big Disasters," *New York Times*, February 20, 2025, https://www.nytimes.com/2025/02/20/climate/trump-cuts-hud-disaster-recovery.html.
5. Karen Zraick, "Farmers Sue over Deletion of Climate Data from Government Websites," *New York Times*, February 24, 2025, https://www.nytimes.com/2025/02/24/climate/agriculture-farmer-website-data-lawsuit.html.
6. Jay Price, "Climate Cutting Initiatives at the DOD Could Come at a Cost to Military Planning," NPR, March 5, 2025, https://www.npr.org/2025/03/05/nx-s1-5316348/cutting-climate-initiatives-at-the-dod-could-come-at-a-cost-to-military-planning.
7. Chelsea Harvey, "Trump's DEI Purge Comes at a Cost to Indigenous Communities," *Scientific American*, February 6, 2025, https://www.scientificamerican.com/article/trumps-dei-purge-comes-at-a-cost-to-indigenous-communities/.
8. "Gowanus Canal Brooklyn, NY Cleanup Activities," Environmental Protection Agency, accessed December 31, 2023, https://cumulis.epa.gov/supercpad/SiteProfiles/index.cfm?fuseaction=second.cleanup&id=0206222.
9. Elizabeth A. Harris, "In Brooklyn, Worrying About Not Only Flooding, but Also What's in Water," *New York Times*, November 5, 2012, https://www.nytimes.com/2012/11/06/nyregion/gowanus-canal-flooding-brings-contamination-concerns.html.

10. *SIRR Analysis* (Sandy Regional Assembly, July 2013), https://nyhealthfoundation. org/wp-content/uploads/2017/11/mayors-special-initiative-rebuilding-resiliency-plan.pdf.

11. "Voices at the People's Climate March: Indigenous Peoples Lead Historic 400,000-Strong People's Climate March," *Cultural Survival Quarterly Magazine*, December 3, 2014, https://www.culturalsurvival.org/publications/cultural-survival-quarterly/voices-peoples-climate-march-indigenous-peoples-lead.

12. Liz Donovan, "New York State Denies Permit for New Astoria Power Plant," *City Limits*, October 27, 2021, https://citylimits.org/2021/10/27/new-york-state-rejects-permit-for-proposed-astoria-power-plant/.

13. Liz Krueger, "Governor Signs Climate Change Superfund Act," New York State Senate, December 26, 2024, https://www.nysenate.gov/newsroom/press-releases/2024/liz-krueger/governor-signs-climate-change-superfund-act.

14. "Accelerating Extinction Rate Triggers Domino Effect of Biodiversity Loss," United Nations, May 21, 2024, https://news.un.org/en/story/2024/05/1150056.

15. Hannah Ritchie and Pablo Rosado, "Fossil fuels are the biggest source of CO2 emissions in most countries, but there are a few exceptions," Our World in Data, November 22, 2024, https://ourworldindata.org/data-insights/fossil-fuels-are-the-biggest-source-of-co2-emissions-in-most-countries-but-there-are-a-few-exceptions.

16. "How Feedback Loops Are Making the Climate Crisis Worse," The Climate Reality Project, January 7, 2020, accessed February 5, 2021, https://climaterealityproject.org/blog/how-feedback-loops-are-making-climate-crisis-worse.

17. Visit the Yale Program on Climate Change Communication for current research on climate attitudes in the United States and other countries: https://climatecommunication.yale.edu.

18. Kabir Agarwal, "World Sees 'Largest Environmental Protest in History' for Climate Action," *Wire*, September 20, 2019, https://thewire.in/environment/climate-change-environmental-protests-india.

19. Eileen Flanagan, "Fighting Fracking in South Africa and Beyond," *Waging Nonviolence*, September 10, 2012, https://wagingnonviolence.org/2012/09/fighting-fracking-in-south-africa-and-beyond/.

20. Zack Budryk, "Trump: 'We want the Keystone XL Pipeline Built!'" *The Hill*, February 25, 2025, https://thehill.com/policy/energy-environment/5162740-trump-we-want-the-keystone-xl-pipeline-built/.

21. Jordan Davidson, "945 Toxic Waste Sites at Risk of Disaster from Climate Crisis," *EcoWatch*, November 18, 2019, https://www.ecowatch.com/superfund-sites-climate-disaster-2641394009.html.

22. Eric W. Sanderson, "Let Water Go Where It Wants to Go," *New York Times*, September 28, 2021, https://www.nytimes.com/2021/09/28/opinion/hurricane-ida-new-york-city.html?fbclid=IwAR31YENIRRMTNLTCsK8Z-x76jlWF8Pd55YEZk vixEbvFuP4OtXl25SjjAyQ.

23. Sherri Mitchell, *Sacred Instructions: Indigenous Wisdom for Living Spirit-Based Change* (North Atlantic Books, 2018), chap. 2, Kindle.

24. Mitchell, *Sacred Instructions*, chap. 18.

25. "When We Love, We Win: Compassionate Activism for November and Beyond—Tara Brach & Friends," streamed live September 21, 2024, by Tara Brach, YouTube, 1 hour, 13 min., 50 sec., https://www.youtube.com/watch?v=ULlıFvrL8rU&t=4644s.

CHAPTER ONE: A CLASH OF WORLDVIEWS

26. V. Masson-Delmotte, et al, eds., *Global Warming of 1.5°C.* (IPCC, 2018), https://www.ipcc.ch/site/assets/uploads/sites/2/2022/06/SR15_Full_Report_HR.pdf.

27. Ingrid Biedron and Suzannah Evans, *Time for Action: Six Years After Deepwater Horizon* (Oceana, April, 2016), https://usa.oceana.org/wp-content/uploads/sites/4/deepwater_horizon_anniversary_report_updated_4-28.pdf.

28. Alyson Flournoy et al., "Regulatory Blowout: How Regulatory Failures Made the BP Disaster Possible, and How the System Can Be Fixed to Avoid a Recurrence," Center for Progressive Reform White Paper no. 1004 (March 2010).

29. Coral Davenport, "Washington Rolls Back Safety Rules Inspired by Deepwater Horizon Disaster," *New York Times*, September 27, 2018, https://www.nytimes.com/2018/09/27/climate/offshore-drilling-safety-deepwater-horizon.html; "Trump rollbacks on fossil fuels will save industry money but cost lives," *Nola*, posted January 27, 2019; "BSEE Proposes Revisions to Production Safety Systems Regulations," Bureau of Safety and Environmental Enforcement, December 28, 2017.

30. Jeffrey A. Groen and Anne E. Polivka, *Hurricane Katrina Evacuees: Who They Are, Where They Are, and How They Are Faring* (US Bureau of Labor Statistics, *Monthly Labor Review*, March 2008), https://www.bls.gov/opub/mlr/2008/03/art3full.pdf.

31. Michael Brown, "Stop Blaming Me for Hurricane Katrina," *Politico Magazine*, August 27, 2015, https://www.politico.com/magazine/story/2015/08/katrina-ten-years-later-michael-brown-121782.

32. Anaïs Teyton and David M. Abramson, "The Formation of Belief: An Examination of Factors That Influence Climate Change Belief Among Hurricane Katrina Survivors," *Environmental Justice* 14, no. 3 (2021): 169–77.

33. Joel Schwarz, "How Media Covered Katrina Aftermath Affects Response by Blacks and Whites," UW News, September 25, 2008, https://www.washington.edu/news/2008/09/25/how-media-covered-katrina-aftermath-affects-response-by-blacks-and-whites-2/.

34. *When the Levees Broke: A Requiem in Four Acts,* directed by Spike Lee, aired August 16, 2006, on HBO.

35. Holly Devon, "Defending the Collective: An Interview with Malik Rahim," *The Iron Lattice*, April 11, 2017, https://theironlattice.com/index.php/2017/04/11/defending-the-collective-an-interview-with-malik-rahim/.

36. Naomi Klein, "How Power Profits from Disaster," *Guardian*, July 6, 2017, https://www.theguardian.com/us-news/2017/jul/06/naomi-klein-how-power-profits-from-disaster; for a more detailed account, see Naomi Klein, *The Shock Doctrine: The Rise of Disaster Capitalism* (Picador, 2008).

37. Beverly Wright, "New Orleans Land Grab: Addressing the 'Elephant' in the City Ten Years After Hurricane Katrina," *Bayview National Black Newspaper*, August 1, 2015, https://sfbayview.com/2015/08/new-orleans-land-grab-addressing-the-elephant-in-the-city-10-years-after-hurricane-katrina/.

38. Rebecca Solnit, "Prelude and Epilogue," in *A Paradise Built in Hell: The Extraordinary Communities That Arise in Disaster* (Penguin Books, 2010), 1, Kindle.

39. George Lakoff, "The Post-Katrina Era," *Huffington Post*, September 8, 2005; updated May 25, 2011, https://www.huffpost.com/entry/the-postkatrina-era_b_7034.

40. George Lakoff, *The Political Mind: Why You Can't Understand 21st-Century American Politics with an 18th-Century Brain* (Viking, 2008), 121.

41. Capitol Lakes: Baton Rouge, Louisiana (EPA National Priorities List, September 2023), https://semspub.epa.gov/work/HQ/404214.pdf.

42. Elizabeth Shogren and Robert Benincasa, "Baton Rouge's Corroded, Overpolluting Neighbor: Exxon Mobil," NPR, May 30, 2013, https://www.npr.org/2013/05/30/187044721/baton-rouge-s-corroded-overpolluting-neighbor-exxon.

43. "Louisiana Refinery Accident Database," Louisiana Bucket Brigade, dataset 2005-2014, https://www.louisianarefineryaccidentdatabase.org.

44. Environmental Integrity Project, *"Breath to the People": Sacred Air and Toxic Pollution* (United Church of Christ, 2020), 22-25, https://www.ucc.org/wp-content/uploads/unitedchurchofchrist/pages/24840/attachments/original/1582721312/FINAL_BreathToThePeople_2.26.2020.pdf?1582721312.

45. Lisa Friedman, "Oil Interests Gave More Than $75 Million to Trump PACs, New Analysis Shows," *New York Times*, November 1, 2024, https://www.nytimes.com/2024/11/01/climate/oil-gas-donations-trump.html.

46. "People Gonna Rise Like the Water: Song Lyrics for the Streets," posted April 14, 2017, by 350.org, YouTube, https://www.youtube.com/watch?v=QKcA9LS8Qpc.

47. Karen Savage, "Sheriff's Deputies Protect Corporate Interests in Bayou Bridge Case," Truthout, December 12, 2018, https://truthout.org/articles/sheriffs-deputies-protect-corporate-interests-in-bayou-bridge-case/; Brett Wilkins, "As DA Rejects Charges, Bayou Bridge Water Protectors Vow 'We Will Not Stop Our Work,'" *Common Dreams*, July 13, 2021, https://www.commondreams.org/news/2021/07/13/da-rejects-charges-bayou-bridge-water-protectors-vow-we-will-not-stop-our-work.

48. Tim Murphy, "A Massive Chemical Plant Is Poised to Wipe This Louisiana Town off the Map," *Mother Jones*, March 27, 2014, https://www.motherjones.com/environment/2014/03/sasol-mossville-louisiana/.

49. Benjamin Franta, "Shell and Exxon's Secret 1980s Climate Change Warnings," *Guardian*, September 19, 2018, https://www.theguardian.com/environment/climate-consensus-97-per-cent/2018/sep/19/shell-and-exxons-secret-1980s-climate-change-warnings.

50. Geoffrey Supran and Naomi Oreskes, "Assessing ExxonMobil's Climate Change Communications (1977–2014)," *IOP Science*, August 23, 2017, https://iopscience.iop.org/article/10.1088/1748-9326/aa815f.

51. Michael Hiltzik, "A New Study Shows How Exxon Mobil Downplayed Climate Change When It Knew the Problem was Real," *Los Angeles Times,* August 22, 2017, https://www.latimes.com/business/hiltzik/la-fi-hiltzik-exxonmobil-20170822-story.html.

52. Mark Kaufman, "The Carbon Footprint Scam," *Mashable,* July 13, 2020, https://mashable.com/feature/carbon-footprint-pr-campaign-sham.

CHAPTER TWO: POWER WITHOUT LOVE

53. Some details from *A Village Called Versailles,* produced and directed by S. Leo Chiang, Walking Iris Films. Also http://docs.virgilhenrystorr.org/chamlee-wrightstorrclubgoods.pdf.

54. Vanessa Hua, "Standing Up for the Vietnamese Community of New Orleans," *Bloomberg CityLab,* October 27, 2014, https://www.bloomberg.com/news/.articles/2014-10-27/standing-up-for-the-vietnamese-community-of-new-orleans.

55. Kevin Litten, "City Council Fines Entergy for Paid Actors Scandal, Affirms Power Plant Vote," *Times-Picayune,* February 22, 2019, https://www.nola.com/news/article_f2266bc6-6458-5b14-87d5-c1e628650210.html.

56. Julie Dermansky, "Entergy Gas Plant Opponents Question Integrity of New Orleans City Council as It Gives Final Approval," *Desmog,* February 27, 2019, https://www.desmogblog.com/2019/02/27/entergy-natural-gas-plant-new-orleans-city-council-approval-lawsuits.

57. Michael Isaac Stein, "Actors Were Paid to Support Entergy's Power Plant at New Orleans City Council Meetings," *Lens,* May 4, 2018, https://thelensnola.org/2018/05/04/actors-were-paid-to-support-entergys-power-plant-at-new-orleans-city-council-meetings/.

58. Jessica Williams, "New Orleans Council Violated Law, but Vote on Entergy Power Plant Still OK, Judges Rule," *Times-Picayune,* February 13, 2020, https://www.nola.com/news/politics/article_13af7070-4e81-11ea-bf49-db5ebb91c620.html.

59. "Martin Luther King Jr., 'Where Do We Go from Here?' FULL SPEECH—August 16, 1967," posted January 22, 2014, by nicholasflyer, YouTube, 37 min., 28 sec. through 39 min., 51 sec., https://www.youtube.com/watch?v=5m1PRN9VCfw&t=9s; see also a fuller explanation in Martin Luther King Jr., *Where Do We Go from Here?* (Beacon Press, 2010), chap. 2, Kindle.

60. Sean Reilly, "Outgoing Air Office Chief's Legacy Tied to Divisive Regs," *Politico,* January 14, 2021, https://www.eenews.net/articles/outgoing-air-office-chiefs-legacy-tied-to-divisive-regs/.

61. "Rev. Manning at New Orleans City Council Meeting," posted March 1, 2019, by Julie Dermansky, YouTube, 47 sec. through 3 min., 33 sec., https://www.youtube.com/watch?v=jbR2BIpw_d4.

62. Sophie Kasakove and Nicholas Bogel-Burroughs, "New Orleans Built a Power Plant to Prepare for Storms. It Sat Dark for 2 Days," *New York Times,* September 10, 2021, https://www.nytimes.com/2021/09/10/us/ida-new-orleans-power.html.

63. Julio Alicea, "African American Passengers Boycott Segregated Buses in Baton Rouge, 1953," Global Nonviolent Action Database, December 9, 2010, https://

nvdatabase.swarthmore.edu/content/african-american-passengers-boycott-segregated-buses-baton-rouge-1953.

64. Eric Douglas, "Coal Production Drop Off Leaves Behind Unreclaimed Mine Lands," West Virginia Public Broadcasting, September 17, 2021, https://wvpublic.org/coal-production-drop-off-leaves-behind-unreclaimed-mine-lands/#:~:text=Coal%20has%20been%20"king"%20for,gas%20have%20reduced%20its%20dominance.

65. "Global Fossil Fuel Divestment Commitments Database," accessed March 7, 2025, https://divestmentdatabase.org.

66. Jon Hale, "Who's Behind the Right's Anti-ESG Campaign?" Morningstar, October 14, 2022, https://www.morningstar.com/sustainable-investing/whos-behind-rights-anti-esg-campaign.

67. Louisiana Bucket Brigade, "Pastor Manning's Arrest," Facebook, January 3, 2020, https://www.facebook.com/watch/?v=1244990205698876; "Why We Have to Speak: Local Pastor Arrested on Final Day of March Against Genocide," *New Orleans Tribune*, November 25, 2019, https://theneworleanstribune.com/2019/11/25/why-we-have-to-speak-local-pastor-arrested-on-final-day-of-march-against-genocide/.

68. Kazu Haga, *Healing Resistance: A Radically Different Response to Harm* (Parallax Press, 2020), 140–41.

69. Matthew Taylor, "The Evolution of Extinction Rebellion," *Guardian*, August 4, 2020, https://www.theguardian.com/environment/2020/aug/04/evolution-of-extinction-rebellion-climate-emergency-protest-coronavirus-pandemic.

70. Nambi Ndugga and Samantha Artiga, "Continued Rises in Extreme Heat and Implications for Health Disparities," KFF, August 24, 2023, https://www.kff.org/racial-equity-and-health-policy/issue-brief/continued-rises-in-extreme-heat-and-implications-for-health-disparities/.

CHAPTER THREE: WHAT'S RACE GOT TO DO WITH IT?

71. This insight came from Anne Rolfes, who led a bike tour of Norco I attended in early March, 2019. Details in this chapter are drawn from Anne's tour, my September 2018 interview with Margie Richard, my February 2018 interview with Wilma Subra, and a variety of written sources, cited when quoted.

72. Steve Lerner, *Diamond: A Struggle for Environmental Justice in Louisiana's Chemical Corridor* (MIT Press, 2005), chap. 3., Kindle.

73. Lerner, *Diamond*, chap. 6.

74. Vann R. Newkirk II, "Trump's EPA Concludes Environmental Racism is Real," *Atlantic*, February 28, 2018, https://www.theatlantic.com/politics/archive/2018/02/the-trump-administration-finds-that-environmental-racism-is-real/554315/.

75. Ibram X. Kendi, *How to Be an Antiracist* (Bodley Head, 2019), 17–18.

76. Margie said this to Anne Rolfes, who quoted it to me on her bike tour in 2019.

77. Lara J. Cushing et al., "Historical Red-Lining is Associated with Fossil Fuel Power Plant Siting and Present-Day Inequalities in Air Pollutant Emissions. *Nature Energy* 8, (2023): 52–61, https://doi.org/10.1038/s41560-022-01162-y.

78. Dorceta E. Taylor, *Toxic Communities: Environmental Racism, Industrial Pollution, and Residential Mobility* (New York University Press, 2014), 29.

79. Ibram X. Kendi, *Stamped from the Beginning: The Definitive History of Racist Ideas in America.* (Bold Type Books, 2017), 29, 98.

80. Lerner, *Diamond*, chap. 6.

81. Taylor, *Toxic Communities*, 105, emphasis in original, and 102.

82. Lerner, *Diamond*, chap. 12.

83. *Toxic Wastes and Race in the United States: A National Report on the Racial and Socio-Economic Characteristics of Communities with Hazardous Waste Sites* (Commission for Racial Justice, United Church of Christ, 1987), https://www.nrc.gov/docs/ML1310/ML13109A339.pdf.

84. "2004 Goldman Prize Winner Margie Richard," The Goldman Environmental Prize, accessed April 16, 2025, https://www.goldmanprize.org/recipient/margie-richard/.

85. Emily Bazelon, "Bad Neighbors," *Legal Affairs*, May–June, 2003, 4, https://www.toxicdocs.org/d/5bE5O4XeDJjDjRKkYo5aZodxJ?lightbox=1# =; see also Lerner, *Diamond*, chap. 6.

86. David Amsden, "Building the First Slavery Museum in America," *New York Times Magazine*, February 26, 2015, https://www.nytimes.com/2015/03/01/magazine/building-the-first-slave-museum-in-america.html.

87. *St. James Parish Government Comprehensive Plan 2031*, (South Central Planning and Development Commission, 2014), 3, https://www.stjamesla.com/DocumentCenter/View/283/St-James-Parish-Comprehensive-Plan-PDF?bidId=.

88. "Margie Richard: 2004 Goldman Prize Winner," posted October 2, 2013, by Goldman Environmental Prize, YouTube, 2 min., 17 sec., to 3 min., 30 sec., https://www.youtube.com/watch?v=j3MaAi1Dl9c; also see Lerner, *Diamond*, chap. 4.

89. Elowyn Corby, "Ogoni people struggle with Shell Oil, Nigeria, 1990-1995," Global Nonviolent Action Database, November 3, 2011, https://nvdatabase.swarthmore.edu/content/ogoni-people-struggle-shell-oil-nigeria-1990-1995; Amnesty International, "Investigate Shell for Complicity in Murder, Rape, and Torture," November 28, 2017, https://www.amnesty.org/en/latest/news/2017/11/investigate-shell-for-complicity-in-murder-rape-and-torture/; "Case Study: Wiwa v. Shell," posted October 23, 2014, by EarthRights International, YouTube, https://www.youtube.com/watch?v=VC9qBPJk_mI.

90. Lerner, *Diamond*, chap. 14.

CHAPTER FOUR: DIVIDING THE SPECTRUM OF ALLIES

91. Environmental Protection Agency, "2022 Greenhouse Gas Emissions from Large Facilities," accessed May 30, 2024, https://ghgdata.epa.gov/ghgp/service/facilityDetail/2022?id=1005823&ds=E&et=&popup=true.

92. "History of Calcasieu Parish," Calcasieu Parish Police Jury, accessed March 10, 2025, https://www.calcasieu.gov/public-visitors/parish-history.

93. Peggy Frankland and Susan Tucker, *Women Pioneers of the Louisiana Environmental Movement* (University Press of Mississippi, 2015), XVIII.

94. Oliver A. Houck, "Willow Springs: a Louisiana Civil Action," *Loyola Law Review* 62 (June 22, 2016), https://law.loyno.edu/sites/law.loyno.edu/files/file_attach/Houck-Final-10-20-16.pdf.

95. Frankland and Tucker, *Women Pioneers*, 44–48.

96. This quote and some details of Peggy's story come from my interview with her on February 18, 2018. Others come from her book *Women Pioneers*.

97. Houck, "Willow Springs," 162.

98. Frankland and Tucker, *Women Pioneers*, 65–69.

99. Arlie Russell Hochschild, *Strangers in Their Own Land: Anger and Mourning on the American Right* (The New Press, 2016), chap. 2-3, Kindle.

100. Daniel Tope, Justin T. Pickett, and Ted Chiricos "Anti-Minority Attitudes and Tea Party Movement Membership," *Social Science Research* 51 (2015): 322–37, https://www.sciencedirect.com/science/article/abs/pii/S0049089X14001793.

101. Riley Snyder, "Reid Right on Claiming Kochs Produce More Pollution than Oil Giants," *Politifact*, July 19, 2016, https://www.politifact.com/factchecks/2016/jul/19/harry-reid/reid-right-claiming-kochs-produce-more-pollution-o/.

102. Leslie M. Harris, *In the Shadow of Slavery: African Americans in New York City, 1626 – 1863* (University of Chicago Press, 2003), 23.

103. Thandeka, *Learning to Be White: Money, Race, and God in America* (Continuum Publishing, 2002), 46–7.

104. Lynn Peril, "Mother Jones Was a Hell-Raising Trailblazer in the Fight For Workers' Rights," *Bust Magazine*, January 30, 2019, https://bust.com/mother-jones-labor-movement/.

105. Calvin Schermerhorn, "The Thibodaux Massacre Left 60 African-Americans Dead and Spelled the End of Unionized Farm Labor in the South for Decades," *Smithsonian*, November 21, 2017, https://www.smithsonianmag.com/history/thibodaux-massacre-left-60-african-americans-dead-and-spelled-end-unionized-farm-labor-south-decades-180967289/.

106. Alan G. Gauthreaux, "An Inhospitable Land: Anti-Italian Sentiment and Violence in Louisiana, 1891–1924," *Louisiana History: The Journal of the Louisiana Historical Association* 51, no. 1 (Winter 2010): 41–68, http://www.jstor.org/stable/40646346.

107. Chris Mahin, "New Orleans, November 1892: One City's Heroic General Strike Defies Racial Divisions," reprinted from Unite Here website by Brotherhood of Maintenance of Way Employes Division, March 7, 2010, https://www.uprfbmwed.org/index.cfm?zone=/unionactive/view_article.cfm&HomeID=95506.

108. *Birds of Prey: Conoco, Condea Vista, and PPG Feeding off of Mossville and Calcasieu Parish* (Louisiana Bucket Brigade), accessed April 16, 2025, 8–9, http://www.pvcinformation.org/assets/pdf/birdsofprey.pdf.

109. Frankland and Tucker, *Women Pioneers*, 81–84.

110. Julie Schwartzwald Meaders, "Health Impacts of Petrochemical Expansion in Louisiana and Realistic Options for Affected Communities," *Tulane Environmental Law Journal* 34, no. 1 (2021): 124–32, https://www.jstor.org/stable/27089955; Heather Rogers, "Erasing Mossville: How Pollution Killed a Louisiana Town," *The Intercept*, November 4, 2015, https://theintercept.com/2015/11/04/erasing-mossville-how-pollution-killed-a-louisiana-town/; Wilma Subra, Mossville Environmental Action

Now, Inc., and Advocates for Environmental Human Rights, *Industrial Sources of Dioxin Poisoning in Mossville, LA: A Report Based on the Government's Own Data*, (2007), 6, https://www.loe.org/images/content/100423/mossville.pdf.

111. James Ridgeway, "Environmental Espionage: Inside a Chemical Company's Louisiana Spy Op," *Mother Jones*, May 20, 2008, https://www.motherjones.com/environment/2008/05/environmental-espionage-inside-chemical-companys-louisiana-spy-op/.

112. A sampling of documents are linked in James Ridgeway, "Black Ops, Green Groups," *Mother Jones*, April 11, 2008, https://www.motherjones.com/environment/2008/04/exclusive-cops-and-former-secret-service-agents-ran-black-ops-green-groups/.

113. *Intelligence Analysis for Dow Global Trends Tracking Team: Activists, Issues and Trends* (Beckett Brown International, August 25, 1999), 4, https://www.motherjones.com/wp-content/uploads/legacy/news/feature/2008/04/Dow-Global-Trends-Tracking-Team.pdf.

114. Kendi, *Stamped*, 465–67, 474–75.

115. Rick Mullin, "Mossville's End," *Chemical & Engineering News*, March 21, 2016, https://cen.acs.org/articles/94/i12/Mossvilles-end.html.

116. Kristin Mosbrucker, "Sasol wraps up $12.8B petrochemical complex in Lake Charles," *Advocate*, November 17, 2020, https://www.theadvocate.com/baton_rouge/news/business/sasol-wraps-up-12-8b-petrochemical-complex-in-lake-charles/article_782c2388-2912-11eb-9ebf-4b50d1e460b5.html.

117. Joanna Macy, "Spiritual Practices for Times of Crisis," *Huffington Post*, June 4, 2011, https://www.huffpost.com/entry/spirituality-crisis_b_871311.

118. adrienne maree brown, "Unthinkable Thoughts: Call Out Culture in the Age of COVID-19," July 17, 2020, https://adriennemareebrown.net/2020/07/17/unthinkable-thoughts-call-out-culture-in-the-age-of-covid-19/, punctuation in the original.

119. Oliver Laughland, "Multibillion-Dollar Louisiana Plastics Plant Put on Pause In a Win for Activists," *Guardian*, August 18, 2021, https://www.theguardian.com/us-news/2021/aug/18/louisiana-plastics-plant-toxic-emissions-cancer-alley?fbclid=IwAR1Tncxj-oY8oUIj9LRz5xEJGDcBeoWx1BJMhJzpVsEWn3GVnK9FUpegXMo.

CHAPTER FIVE: NAVIGATING THE CURRENTS OF RACE

120. Visit the Choose Democracy website for resources on what to do to protect democracy in the United States, https://choosedemocracy.us.

121. George Lakey, *How We Win: A Guide to Nonviolent Direct Action Campaigning* (Melville House, 2018), 33–34.

122. Paul Solman, "This La. Battle is Between Big Industry and a Green Army," PBS, May 30, 2019, https://www.pbs.org/newshour/show/this-la-battle-is-between-big-industry-and-a-green-army.

123. Julie Dermansky, "Bayou Corne Residents Still Evacuated Nearly Two Years After Salt Mine Collapse Caused Sinkhole," *Truthout*, May 13, 2014, https://truthout.

org/articles/bayou-corne-residents-still-evacuated-nearly-two-years-after-salt-mine-collapse-caused-sinkhole/.

124. Miranda Fox, "City Council Must Hold Entergy Accountable for Failing New Orleans Again," Earthjustice, September 14, 2021, https://earthjustice.org/news/press/2021/city-council-must-hold-entergy-accountable-for-failing-new-orleans-again.

125. Geoff Dembicki, "A Debate over Racism Has Split One of the World's Most Famous Climate Groups," *VICE*, April 28, 2020, https://www.vice.com/en/article/jgey8k/a-debate-over-racism-has-split-one-of-the-worlds-most-famous-climate-groups.

126. "Extinction Rebellion America Co-Founder Jonathan Logan Leads Change," Powerhouse Leaders Podcast, March 5, 2020, 5 min., 41 sec., https://www.youtube.com/watch?v=kix_S1PpuDM.

127. "(1977) The Combahee River Collective Statement," BlackPast, November 16, 2012, https://www.blackpast.org/african-american-history/combahee-river-collective-statement-1977/.

128. Keeanga-Yamahtta Taylor, ed., *How We Get Free: Black Feminism and the Combahee River Collective* (Haymarket Books, 2017), 11.

129. Heather McGhee, *The Sum of Us: What Racism Costs Everyone and How We Can Prosper Together* (One World, 2021).

130. Claire Schaeffer-Duffy, "Sharon Lavigne's Fighting Faith on the Bayou," *EarthBeat*, a project of *National Catholic Reporter*, October 30, 2020, https://www.ncronline.org/news/earthbeat/sharon-lavignes-fighting-faith-bayou.

131. AP, "Juneteenth at Slave Cemetery on $9.4B Construction Site," *Taiwan News*, June 20, 2020, https://www.taiwannews.com.tw/en/news/3951969; "RISE St. James Attorneys Speak After a Victory in Court Against Formosa on June 18," posted on June 19, 2020, by Julie Dermansky, YouTube, 18 sec., https://www.youtube.com/watch?v=Pv_sJUrFWK0.

132. "The Current: The Race/Class Narrative," City Club of Portland, streamed live on October 9, 2019, by Anat Shenker-Osorio, YouTube, 12 min., 40 sec., https://www.youtube.com/watch?v=8hQo3UhOFpA.

CHAPTER SIX: ALL OUR RELATIONS

133. Robin Wall Kimmerer, *Braiding Sweetgrass: Indigenous Wisdom, Scientific Knowledge, and the Teachings of Plants* (Milkweed Editions, 2013), 368–69.

134. Toban Black et al., eds., *A Line in the Tar Sands: Struggles for Environmental Justice* (PM Press, 2014).

135. James Hansen, "Game Over for the Climate," *New York Times*, May 9, 2012, https://www.nytimes.com/2012/05/10/opinion/game-over-for-the-climate.html; David Biello, "Greenhouse Goo," *Scientific American* 308, no. 7 (2013): 56–61.

136. Roxanne Dunbar-Ortiz, *An Indigenous Peoples' History of the United States* (Beacon Press, 2014), 157–61; see also, "Relations: Dakota and Ojibwe Treaties," Why Treaties Matter, accessed April 17, 2025, https://treatiesmatter.org/relationships/basis-of-civilization.

137. Anton Treuer and Tom Weber, "Warrior Nation: A History of the Red Lake Ojibwe," Minnesota Public Radio, October 19, 2015, https://www.mprnews.org/story/2015/10/19/bcst-warrior-nation-a-history-of-the-red-lake-ojibwe.

138. Kimmerer, *Braiding Sweetgrass*, 365–68.

139. Joanna Macy, "Spiritual Practices for Times of Crisis," *Huffington Post*, June 4, 2011, https://www.huffpost.com/entry/spirituality-crisis_b_871311.

140. Winona LaDuke, "The Indigenous Growers Reviving Hemp's Deep Roots," MSN, May 20, 2024, https://www.msn.com/en-us/society-culture-and-history/general/the-indigenous-growers-reviving-hemp-s-deep-roots/ar-BB1mJ8kY.

141. Nina Lakhani, "FBI's Opposition to Releasing Leonard Peltier Driven by Vendetta, Says Ex-Agent," *Guardian*, January, 18, 2023, https://www.theguardian.com/us-news/2023/jan/18/leonard-peltier-clemency-fbi-agent-coleen-rowley; "UN Body Finds Activist's Detention 'Arbitrary' in Case Filed by Lowenstein Clinic," Yale Law School, August 8, 2022, https://law.yale.edu/yls-today/news/un-body-finds-activists-detention-arbitrary-case-filed-lowenstein-clinic.

142. Evan Simon et al., "Pipeline Firm Deposited Millions into State Fund to Pay Local Police to 'Patrol' and 'Protect' Controversial Line 3 Project," ABC News, November 1, 2021, https://abcnews.go.com/US/pipeline-firm-deposited-millions-state-fund-pay-local/story?id=80844727.

143. Alex Binder, "Enbridge Spills 10,000 Gallons of Line 3 Drilling Fluid," *Unicorn Riot*, August 16, 2021, https://unicornriot.ninja/2021/enbridge-spills-10000-gallons-of-line-3-drilling-fluid/?fbclid=IwAR2CiAMoAItBtNKJh9uiU8ZxbLYdTLob4r4ihHeWmmxwOor3I-_-dIq-py4; Aki Nace and Erin Hassanzadeh, "Line 3 Oil Pipeline: A Look at What's Happened Since the Pipeline Started Operating in Northern Minnesota," CBS, May 3, 2023, https://www.cbsnews.com/minnesota/news/line-3-oil-pipeline/.

144. Kimmerer, *Braiding Sweetgrass*, 369.

145. Kimmerer, *Braiding Sweetgrass*, 3–10.

146. Trista Lara, "The US Government's Legislative Shortcomings Are to Blame for Murdered Missing Indigenous Women," *New University*, May 23, 2023, https://newuniversity.org/2023/05/23/the-u-s-governments-legislative-shortcomings-are-to-blame-for-murdered-missing-indigenous-women/.

147. "Religious Crimes Code of 1883 Bans Native Dances, Ceremonies," Investing in Native Communities, accessed March 11, 2025, https://nativephilanthropy.candid.org/events/religious-crimes-code-of-1883-bans-native-dances-ceremonies/.

148. Randy Furst, "Judge Dismissed Charges Against Activists Accused of Disrupting Enbridge Line 3," *Minnesota Star Tribune*, September 18, 2023, https://www.startribune.com/judge-dismisses-charges-against-activists-accused-of-disrupting-enbridge-line-3/600305634.

149. "'I'm Not a Criminal . . . Enbridge Is': Charges Tossed Against Winona LaDuke & Others for Pipeline Action," *Democracy Now!*, September 30, 2023, https://www.democracynow.org/2023/9/20/winona_laduke_line_3_charges.

150. Mark Engler and Paul Engler, *This Is an Uprising: How Nonviolent Revolt Is Shaping the Twenty-First Century* (Nation Books/Perseus Books Group, 2016), 80.

CHAPTER SEVEN: JUST TRANSITION

151. Paula Giese, "Navajo-Hopi Long Land Dispute," 1996, last updated March 20, 1997, http://www.kstrom.net/isk/maps/az/navhopi.html.

152. Enei Begaye, "The Black Mesa Controversy," Cultural Survival, May 7, 2010, https://www.culturalsurvival.org/publications/cultural-survival-quarterly/black-mesa-controversy.

153. Jessica Kutz, "The Fight for an Equitable Energy Economy for the Navajo Nation," *High Country News*, February 21, 2021, https://www.hcn.org/issues/53.2/south-coal-the-fight-for-an-equitable-energy-economy-for-the-navajo-nation.

154. Jeff Biggers, "Green Bounty: Historic Navajo Green Jobs Legislation," *Grist*, July 23, 2009, https://grist.org/climate-energy/green-bounty-historic-navajo-green-jobs-legislation/.

155. "Wahleah Johns: Executive Director Black Mesa Water Coalition," Women in Green Forum, September 24, 2011, 3 min., 58 sec., https://www.youtube.com/watch?v=ej8PKNd_B1Y.

156. Kutz, "Fight for an Equitable Energy."

157. "Testimony of Nicole Horseherder," Committee on Natural Resources, June 4, 2012, https://naturalresources.house.gov/uploadedfiles/horseherdertestimony06-04-12.pdf.

158. Bobby Magill, "The West's Largest Coal Plant—and Seventh Biggest Source of CO_2 Emissions in the US—May Close," *Salon*, February 2, 2017, https://www.salon.com/2017/02/02/the-wests-largest-coal-plant-and-seventh-biggest-source-of-co2-emissions-in-the-u-s-may-close/.

159. "Testimony of Bidtah N. Becker, Legal Council," Subcommittee on Indian and Insular Affairs, September 28, 2023, https://docs.house.gov/meetings/II/II24/20230928/116420/HHRG-118-II24-Wstate-BeckerB-20230928.pdf.

160. Kutz, "Fight for an Equitable Energy."

161. My definition is informed by the summary of Robert Blauner's ideas in Dorceta Taylor's *Toxic Communities*, 47–48.

162. "Columbus reports on his first voyage, 1493," *The Gilder Lehrman Institute of American History AP US History Guide*, https://ap.gilderlehrman.org/resource/columbus-reports-his-first-voyage-1493?period=1.

163. Steven T. Newcomb, *Pagans in the Promised Land: Decoding the Doctrine of Christian Discovery* (Chicago Review Press, 2008); see original text: "The Papal Bull *Inter Caetera* of May 4, 1493," introduced and translated by Sebastian Modrow and Melissa Smith, Doctrine of Discovery Project, accessed April 17, 2025, https://doctrineofdiscovery.org/assets/pdfs/Inter_Caetera_Modrow&Smith.pdf.

164. Garrick Bailey and Roberta Glenn Bailey, *A History of the Navajos: The Reservation Years* (School of American Research Press, 1986).

165. The detail about the snakes came from a talk at the Canyon de Chelly Visitor Center, June 2018.

166. "Treaty Between the United States and the Navajo Indians Signed at Fort Sumner, N.M., June 1, 1868," Bosque Redondo Memorial Digital Collections, accessed

March 20, 2025, https://archive-bosqueredondomemorial.nmhistoricsites.org/scripto/1/12723/12732, emphasis added.

167. Bailey and Bailey, *History of the Navajos*, 111.

168. Jared Farmer, "Erosion of Trust," *American Scientist* 98, no. 4 (July–August, 2010): 348, https://www.americanscientist.org/article/erosion-of-trust; "Boulder Dam and the Navajo Reservation," *Native American Roots*, November 11, 2014, http://nativeamericannetroots.net/diary/1761.

169. Marsha Weisiger, "Gendered Injustice: Navajo Livestock Reduction in the New Deal Era," *Western Historical Quarterly* 38, no. 4 (2007): 437–55.

170. Eric Henderson, "Navajo Livestock Wealth and the Effects of the Stock Reduction Program of the 1930s," *Journal of Anthropological Research* 45, no. 4 (1989): 379–403.

171. Donald A. Grinde Jr., "Navajo Opposition to the Indian New Deal," *Equity & Excellence in Education* 19, no. 3-6 (2010): 79–87.

172. Bailey and Bailey, *History of the Navajo*, 161.

173. John Dougherty, "A People Betrayed," *Phoenix New Times*, May 1, 1977, https://www.phoenixnewtimes.com/news/a-people-betrayed-6423155.

174. "Broken Rainbow—Part 6 of 7," posted November 5, 2008, by Morningwasichu, YouTube, 9 sec., https://www.youtube.com/watch?v=-9LctAViabA/.

175. Alan Ahn, Matt Bowen, and Sagatom Saha, "Achieving Critical Mass: Trump Administration Policies that Would Strengthen Nuclear Energy," MSN, February 23, 2025, https://www.msn.com/en-us/news/us/achieving-critical-mass-trump-administration-policies-that-would-strengthen-nuclear-energy/ar-AA1ylyyB.

176. Judy Pasternak, *Yellow Dirt: A Poisoned Land and the Betrayal of the Navajos* (Free Press, 2010).

177. Will Ford, "A Radioactive Legacy Haunts This Navajo Village, Which Wears a Fractured Future," *Washington Post*, January 18, 2020, https://www.washingtonpost.com/national/a-radioactive-legacy-haunts-this-navajo-village-which-fears-a-fractured-future/2020/01/18/84c6066e-37e0-11ea-9541-9107303481a4_story.html#comments-wrapper.

178. Chris Shuey and Johnnye Lewis, *Preliminary Results of the Navajo Birth Cohort Study and Selected Case Studies of Exposures to Uranium in Mining Wastes and Drinking Water* (Navajo Department of Public Health December 3, 2015), presented at the Annual Meeting of the CHR Outreach Program, http://sric.org/nbcs/docs/NDOH_CHR_conf_presentation_120315.pdf.

179. Francesca Benson, "The Largest Accidental Release of Radioactive Material in US History: What Happened at Church Rock?," IFLScience, April 19, 2022; "40th Anniversary of the Church Rock Spill, Part 02," Chris Shuey speaking, July 13, 2009, 2 min., 13 sec., https://www.youtube.com/watch?v=-__kcw9QoYA.

180. Ford, "Radioactive Legacy," and Pasternak, *Yellow Dirt*, 148–150.

181. AP, "14 Year Cleanup at Three Mile Island Concludes," *New York Times*, August 15, 1993, https://www.nytimes.com/1993/08/15/us/14-year-cleanup-at-three-mile-island-concludes.html; and "Backgrounder on the Three Mile Island Accident," USNRC, updated March 28, 2024, https://www.nrc.gov/reading-rm/doc-collections/fact-sheets/3mile-isle.html.

182. "A Case Study of Chronic Uranium Exposure in Sanders, AZ," Northern Arizona University, November 21, 2017, https://mediaspace.nau.edu/media/rt+2+of+3A+A+Case+Study+of+Chronic+Uraniurs%2C+AZ/0_a4qx4fca/191147813.

183. Julia Lurie, "Small-Town America Has a Serious Drinking-Water Problem," *Mother Jones*, June, 2016, https://www.motherjones.com/environment/2016/06/rural-water-contamination-uranium-navajo-sanders/.

184. Ryan Heinsius, "Indigenous Leaders and Activists Protest Uranium Shipments Across Navajo Nation," NPR, August 7, 2024, https://www.npr.org/2024/08/07/nx-s1-5061780/indigenous-leaders-and-activists-protest-uranium-shipments-across-navajo-nation; Gabriel Pietrorazio, "Along a 300-mile Hauling Route, Navajo Uranium Concerns Collide with US Clean Energy Agenda," KSJD, July 11, 2024, https://www.ksjd.org/2024-07-11/along-a-300-mile-hauling-route-navajo-uranium-concerns-collide-with-u-s-clean-energy-agenda.

185. Arlyssa D. Becenti, "As Uranium Ore Ships Across Navajo Nation, an Agreement Faces Scrutiny," *USA Today*, February 14, 2025, https://www.usatoday.com/story/news/local/arizona-environment/2025/02/14/navajo-leaders-question-uranium-ore-shipping-agreement-terms/78489940007/.

186. Bill Corcoran, "The Sierra Club's Shadowy History With the Navajo Generating Station," *Sierra*, October 12, 2017, https://www.sierraclub.org/sierra/sierra-club-s-shadowy-history-navajo-generating-station; John Redhouse, "Red Truth, White Cover-Up," Censored News, April 2, 2012, https://bsnorrell.blogspot.com/2012/04/john-redhouse-red-truth-whte-cover-up.html.

187. Evelyn Nieves, "The Largest Coal-Fired Power Plant in the West Is Slated for Closure," *Sierra*, October 12, 2017, https://www.sierraclub.org/sierra/2017-6-november-december/feature/largest-coal-fired-power-plant-west-slated-for-closure.

188. Philip G. Terrie, *Contested Terrain: A New History of Nature and People in the Adirondacks* (Syracuse University Press, 2008).

189. Interview with Dan Josephson, Cornell University's Adirondack Fishery Research Program, September 4, 2018; see Debbie Pastore, "Conservationist of the Year 2018," Adirondack Council, June 6, 2018, https://www.adirondackcouncil.org/page/blog-139/news/conservationist-of-the-year-2018--celebrating-science-and-conservation-1089.html.

190. Robert S. McPherson, "Navajo Livestock Reduction in Southeastern Utah, 1933–46: History Repeats Itself," *American Indian Quarterly* 22, no. 1/2 (Winter–Spring, 1998): 1–18; and Grinde, "Navajo Opposition."

191. Pathfinder, "Defend Black Mesa Sovereignty: Protect Diné Ways of Life," Black Mesa Indigenous Support, April 30, 2018, https://supportblackmesa.org/2018/04/defend-black-mesa-sovereignty-protect-dine-ways-of-life-may-14th-2018/.

192. Kutz, "Fight for an Equitable Energy."

193. Within weeks of Trump's second inauguration, his administration tried to renege on clean energy grants, including $1.5 billion to Indigenous communities, though legal challenges soon followed. Brian Edwards, "EPA Moves to Seize $20B in Clean Energy Grants, Including $1.5B for Native Communities," *Tribal Business News*, February 14, 2025, https://tribalbusinessnews.com/sections/energy/15019-

epa-moves-to-seize-20b-in-clean-energy-grants-including-1-5b-for-native-communities.

194. "Yale Climate Opinion Maps 2024," Yale Program on Climate Communication, March 13, 2025, https://climatecommunication. yale.edu/visualizations-data/ycom-us-2024/?utm_ source=Yale+Program+on+Climate+Change+Communication&utm_ campaign=3a44373b51-EMAIL_CAMPAIGN_2025_03_11_05_58&utm_ medium=email&utm_term=0_-3a44373b51-711524172.

195. Alex Hager, "Federal energy regulators deny permits for a controversial project on the Navajo Nation," KUNC NPR, February 16, 2024, https://www.kunc. org/news/2024-02-16/federal-energy-regulators-deny-permits-for-a-controversial-project-on-the-navajo-nation.

196. Denis Payre, "FERC's Decision Wasn't Just a Rejection of PPA, It Was a Shift In Policy," *Navajo Times*, February 29, 20124, https://navajotimes.com/opinion/fercs-decision-wasnt-just-a-rejection-of-ppa-it-was-a-shift-in-policy/.

197. Joshua J. Mark, "Doctrine of Discovery," *World History Encyclopedia*, October 11, 2023, https://www.worldhistory.org/Doctrine_of_Discovery/.

198. "3000-Year-Old Solutions to Modern Problems," posted September 29, 2022, by Lyla June, YouTube, TEDxKC, https://www.youtube.com/ watch?v=eH5zJxQETl4.

199. Mark Clatterbuck, ed., *Sacred Resistance: Eco-Activism and the Rise of New Spiritual Communities* (Orbis Books, 2025).

200. "Case Study & Dialogue II: Oneida Women Return to their Homeland," posted February 10, 2023, by Land Justice Futures, Michelle Schenandoah speaking with Lesili Haines, 8 min., 17 sec., https://www.youtube.com/ watch?v=fUBN4aGSqVE&t=4749s.

201. "Case Study & Dialogue II: Oneida Women."

202. Casey Kleczek, "Saved by Seaweed: Nuns and Native Women Heal Polluted New York Waters Using Kelp," *Guardian,* June 27, 2023, https://www.theguardian.com/ us-news/2023/jun/27/shinnecock-tribe-nuns-kelp-farm-long-island-bay?ref=biztoc. com.

203. "We move together: Shinnecock women & Sisters of St. Joseph," FaithInvest, May 26, 2022, https://www.faithinvest.org/post/we-move-together-shinnecock-women-sisters-of-st-joseph.

204. Pope Francis, Encyclical letter, "Laudato Si': On Care for Our Common Home" (May 24, 2015).

205. James Bruggers, "Catholic Bishops in the US Largely Ignore the Pope's Concern About Climate Change, a New Study Finds," *Inside Climate News*, October 26, 2021, https://insideclimatenews.org/news/26102021/catholic-bishops-pope-francis-climate-change-laudato-si/#:~:text="It's%20widely%20known%20that%20the%20 Bishops%20have,climate%20change%20among%20other%20issues%2C"%20 she%20said.

206. Brian Roewe, "Chicago Archdiocese to Power Parishes, Schools With 100% Renewable Energy," *EarthBeat*, December 21, 2023, https://www.ncronline.org/ earthbeat/faith/chicago-archdiocese-power-parishes-schools-100-renewable-energy.

CHAPTER EIGHT: ONENESS AND DIFFERENCE GLOBALLY

207. Hannah Ritchie, "Who Has Contributed Most to Global CO2 Emissions?" Our World in Data, October 1, 2019, https://ourworldindata.org/contributed-most-global-co2.

208. "United States Prevails in WTO Dispute Challenging India's Discrimination Against US Solar Exports," Office of the United States Trade Representative, February 24, 2016, https://ustr.gov/about-us/policy-offices/press-office/press-releases/2016/february/united-states-prevails-wto-dispute.

209. Sandra Halperin, "Neocolonialism," *Encyclopedia Britannica*, April 5, 2024; updated January 25, 2025, https://www.britannica.com/topic/neocolonialism.

210. Priti David, "Battle of the Bugs: On Wings of Climate Change," PARI (People's Archive of Rural India), September 22, 2020, https://ruralindiaonline.org/en/articles/battle-of-the-bugs-on-wings-of-climate-change/.

211. Climate Impact Lab, *Climate Change and Heat-Induced Mortality in India*, Tata Center for Development at UChicago, November 2019, https://impactlab.org/wp-content/uploads/2019/10/IndiaMortality_webv2.pdf.

212. Emily Schmall and Jack Ewing, "India's Electric Vehicle Push Is Riding on Mopeds and Rickshaws," *New York Times*, September 4, 2022, https://www.nytimes.com/2022/09/04/business/energy-environment/india-electric-vehicles-moped-rickshaw.html.

213. Kieran Mulvaney, "Climate Change Report Card: These Countries Are Reaching Targets," *National Geographic*, September 19, 2019, https://www.nationalgeographic.com/environment/2019/09/climate-change-report-card-co2-emissions/.

214. Akshat Rathi and Archana Chaudhary, "The Poorest Super-Emitter Needs a Different Path to Net Zero," *Bloomberg*, March 18, 2022, https://www.bloomberg.com/news/features/2022-03-19/how-india-gets-to-net-zero.

215. This saying is a reference to G. Ananthapadmanabhan, K. Srinivas, and Vinuta Gopal, "Hiding Behind the Poor," Greenpeace India, October 2007, https://wayback.archive-it.org/9650/20200501143324/http://p3-raw.greenpeace.org/india/Global/india/report/2007/11/hiding-behind-the-poor.pdf.

216. The average carbon footprint of Americans is seven times that of Indians. If we account for longer lifespans and class differences, my children could easily have a carbon footprint twenty times the size of a poorer Indian. Leondro Vigna and Johannes Friedrich, "9 Charts Explain Per Capita Greenhouse Gas Emissions by Country," World Resources Institute, May 8, 2023, https://www.wri.org/insights/charts-explain-per-capita-greenhouse-gas-emissions#:~:text=But%20among%20the%20top%2010,just%202.5%20tCO2e%20per%20person.

217. Priya Pillai et al., "Singrauli: The Coal Curse," Greenpeace India Society, September 2011, https://wayback.archive-it.org/9650/20200401214557/http://p3-raw.greenpeace.org/india/Global/india/report/Fact-finding-report-Singrauli-Report.pdf.

218. "Mahan Forests Inviolate—No Go For Mining Says MoEF" Greenpeace India, February 24, 2015, https://www.greenpeace.org/india/en/press/2353/mahan-forests-inviolate-no-go-for-mining-says-moef/.

219. Ruchira Talukdar, "Democracy Zindabad! A Day in the Life of Anti-Coal Resistance in India's Energy Capital," *New Matilda*, February 10, 2018, https://newmatilda.com/2018/02/10/democracy-zindabad-day-life-anti-coal-resistance-indias-energy-capital/.

220. Samanth Subramanian, "India's War on Greenpeace," *Guardian*, August 11, 2015, https://www.theguardian.com/world/2015/aug/11/indias-war-on-greenpeace.

221. Megan Darby, "Coal Mining Banned in India's Mahan Forest," *Climate Home News*, March 24, 2015, https://www.climatechangenews.com/2015/03/24/coal-mining-banned-in-indias-mahan-forest/.

222. Justin Rowlatt, "Why India's Government is Targeting Greenpeace," BBC, May 16, 2015, https://www.bbc.com/news/world-asia-india-32747649.

223. "Africa Faces Disproportionate Burden From Climate Change and Adaptation Costs," World Meteorological Association, September 2, 2024, https://wmo.int/news/media-centre/africa-faces-disproportionate-burden-from-climate-change-and-adaptation-costs.

224. Eileen Flanagan, *Renewable: One Woman's Search for Simplicity, Faithfulness, and Hope* (She Writes Press, 2015).

225. Joanna Macy quoted on "Earth Empathy," Institute for Earth Regenerative Studies, accessed April 19, 2025, http://www.earthregenerative.org/earth-empathy/hope-cat.html.

226. "COP27 Climate Summit," *New York Times*, November 11, 2022, section by Abdi Latif Dahir, "'A crisis in empathy,': Activists Lament Lack of Solidarity at Climate Talks," https://www.nytimes.com/live/2022/11/11/climate/cop27-climate-summit?algo=clicks_norm_diversified&block=1&campaign_id=142&emc=edit_fory_20221111&fellback=true&imp_id=886696596&instance_id=77352&nl=for-you&nlid=181042087&rank=3&regi_id=181042087&req_id=512421095&segment_id=112936&surface=for-you-email-wym&user_id=d5f3a9d34ef03a324ac2e887e619bed6&variant=0_combo_lda_channelsize5_unique_edimp_fye_step50_diversified#the-us-delegation-came-ready-to-showcase-bidens-clean-energy-act-but-the-reception-is-thats-all.

227. Associated Press, "In Another Climate and Money Withdrawal, US Pulls Out of Climate Damage and Compensation Fund," AP News, March 10, 2025, https://apnews.com/article/trump-climate-compensation-fund-withdrawal-0a74bfb4c2b82ed2748f890bb5b604e5.

228. Ayaz Gul, "Study: Pakistan Flood Damages, Economic Losses Exceed $30 Billion," *Voice of America*, October 28, 2022, https://www.voanews.com/a/study-pakistan-flood-damages-economic-losses-exceed-30-billion-/6810207.html.

229. Oliver Franklin Wallis, "Inside India's Gargantuan Mission to Clean the Ganges River," *Wired*, November 30, 2023, https://www.wired.com/story/india-ganges-river-clean-project/; "The Toxic Cost of Kanpur's Leather Industry," *India Today*, July 21, 2016, https://www.indiatoday.in/fyi/story/the-toxic-cost-of-kanpurs-leather-industry-329990-2016-07-19.

230. Stuart Butler, "The Ganges: River of Life, Religion, and Pollution," *Geographical*, January 20, 2022, https://geographical.co.uk/culture/the-ganges-river-of-life-religion-and-pollution.

CHAPTER NINE: LESSONS FROM THE COVID-19 PANDEMIC

231. Jennifer B. Nuzzo and Jorge R. Ledesma, "Why Did the Best Prepared Country in the World Fare So Poorly during COVID?" *Journal of Economic Perspectives*, 37, no. 4: 3–22, https://pubs.aeaweb.org/doi/pdfplus/10.1257/jep.37.4.3.
232. Eileen Flanagan, "Water Protectors Show that 'Another World is Possible'— Through Resistance and Care," *Waging Nonviolence*, October 21, 2021, https://wagingnonviolence.org/2021/10/people-vs-fossil-fuels-bureau-indian-affairs-protest-water-protectors/.

CHAPTER TEN: APPLYING THE LESSONS IN THE VANGUARD CAMPAIGN

233. Chris Flood, "Vanguard Tops List of World's Largest Coal Investors," *Financial Times*, February 25, 2021, https://www.ft.com/content/c032f5a4-9801-47d3-abf7-d5d9826558cc; to look up recent stats on US companies, enter them in Yahoo Finance and then click "holders."
234. "March 9: Chester Residents and CRCQL Allies Attend the Solid Waste Authority Board Meeting," Chester Residents Concerned for Quality Living, March 9, 2022, https://chesterpaej.org/march-9-chester-residents-and-crcql-allies-attend-the-solid-waste-authority-board-meeting/.
235. "The Freedom Singers—Woke Up This Morning," recorded live at the Newport Folk Festival, July 26, 1963, posted July 14, 2014, https://www.youtube.com/watch?v=TsziXdKfOsE.
236. George Lakey, *Dancing with History: A Life for Peace and Justice* (Seven Stories Press, 2022), 95–107.
237. US Census Bureau (2023), American Community Survey 5-year estimates, retrieved from Census Reporter Profile page for Chester, PA, https://censusreporter.org/profiles/16000US4213208-chester-pa/
238. Dennis Coker, Opening Plenary, "Quakers, Colonization, and Decolonization," Friends Association of Higher Education, Haverford College, June 12, 2023, https://drive.google.com/file/d/1eVCoN8ILE305w6knvSFofvgipvnWPIkl/view.
239. "Pennsylvania Charter to William Penn - March 4, 1681," Pennsylvania Historical and Museum Commission, August 26, 2015, version, accessed April 19, 2025, https://www.phmc.state.pa.us/portal/communities/documents/1681-1776/pennsylvania-charter.html.
240. Quoted in Jean R. Soderlund, "You Are Our Brothers," *Friends Journal*, August 1, 2024, https://www.friendsjournal.org/you-are-our-brothers/.
241. Catherine Dunn, "Walk for Change: Campaign Urges Vanguard to Invest Climate-Friendly," *Philadelphia Inquirer*, April 18, 2022, https://eedition.inquirer.com/infinity/article_popover_share.aspx?guid=dd1a4825-4698-4f58-86a6-f931474141eb&share=true.

242. Sophia Schmidt, "D.C. Is Sending $30M to Fast-Track Cleanup of Soil in Philly," Plan Philly, December 17, 2021, https://whyy.org/articles/d-c-is-sending-30m-to-fast-track-cleanup-of-toxic-soil-in-philly/.

243. "Vanguard's Approach to Climate Risk," Vanguard, 1995–2025, accessed April 19, 2025, https://corporate.vanguard.com/content/corporatesite/us/en/corp/climate-change.html.

244. John Woolman, "A Plea for the Poor," part IX, chap. 10, originally published 1793 , accessed April 19, 2025, https://columbiafriendsmeeting.org/wp-content/uploads/2020/08/john_woolman-a_plea_for_the_poor.pdf. See also Phillip P. Moulton, ed., *The Journal and Major Essays of John Woolman* (Oxford University Press, 1971).

245. Thanks to Quaker Earthcare Witness for organizing the Zoom worship and for publishing an earlier version of this story. Eileen Flanagan, "Following Spirit, Despite Fear: Remembering John Woolman in the Vanguard Campaign," Quaker Earthcare Witness, August 7, 2022, https://quakerearthcare.org/following-spirit-despite-fear-remembering-john-woolman-in-the-vanguard-campaign/.

246. Zoë Schlanger, "America's New Climate Delusion," *Atlantic*, September 3, 2024, https://www.theatlantic.com/science/archive/2024/09/louisiana-climate-carbon-capture-lng/679664/?gift=1wJJOWpbGcy0FRPza_6RtN5tUznHt zAfoc-PdfBRBRQ&utm_source=copy-link&utm_medium=social&utm_campaign=share

247. "Understanding the carbon impacts of Waste to Energy," Zero Waste Europe, March 18, 2020, https://zerowasteeurope.eu/2020/03/understanding-the-carbon-impacts-of-waste-to-energy/.

248. *Reimagine Your World: Sustainability Report 2024* (Reworld, 2024), https://4944195.fs1.hubspotusercontent-na1.net/hubfs/4944195/Reworld%202024%20Sustainability%20Report.pdf?__hstc=197568432.0c1daf5c09f62bc47d8fddac726089783500.1&__hssc=197568432.8.1726089783500&__hsfp=3017609856&hsCtaTracking=8068e6dc-4b4c-4feb-8f13-203d92036ddf%7C52aa6fcf-0ba9-4cbd-9c23-ba21a42dcf97.

249. Environmental Paper Network, Colectivo Viento Sur, and Global Forest Network, "Conflict Plantations, Chapter 3: Stolen Land and Fading Forests in Chile," May 2022, Environmental Paper Network, https://environmentalpaper.org/wp-content/uploads/2022/06/20220530-Arauco-en.pdf; Sergio Baffoni, "Indigenous Mapuche Forest Activist Murdered in Chile by Police," Environmental Paper Network, June 30, 2020, https://environmentalpaper.org/2020/06/indigenous-mapuche-forest-activist-murdered-in-chile-by-police/.

250. "ESG Investing," Vanguard, accessed September 13, 2024, https://investor.vanguard.com/investment-products/esg; "Mutual Funds and ETFs," Fossil Free Funds, accessed September 13, 2024, https://fossilfreefunds.org/funds?q=Vanguard.

251. David Bach, "The American Right Has Gone to War With 'Woke Capitalism'—Here's What They Get Wrong," *Conversation*, February 24, 2023, https://theconversation.com/the-american-right-has-gone-to-war-with-woke-capitalism-heres-what-they-get-wrong-200580.

252. Jessica Guynn, "GOP vs. ESG: Why Florida Gov. Ron DeSantis, Republicans are Fighting 'Woke' ESG Investing," *USA Today*, December 19, 2022, https://www.usatoday.com/story/money/2022/12/19/what-are-esg-investments-businesses/10898841002/.

253. George Lakey, "How to Build a Progressive Movement in a Polarized Country," *Waging Nonviolence*, March 16, 2018, https://wagingnonviolence.org/2018/03/how-build-progressive-movement-polarized-country/.

254. Ryan Phillips, "How Private Equity Firms Caused the UMWA Strike in Brookwood," February 18, 2022, Patch, https://patch.com/alabama/tuscaloosa/how-private-equity-firms-caused-umwa-strike-brookwood.

255. "Historic Hanford Contamination is Worse than Expected: Oregon Experts Weigh in," Oregon Department of Energy, July 7, 2023, https://energyinfo.oregon.gov/blog/2023/7/7/historic-hanford-contamination-is-worse-than-expected-oregon-experts-weigh-in.

256. Because of concern about people trying to copy or exploit this sacred tradition, Gigi said I shouldn't describe the ceremony itself, but that I could share my own experience of it.

257. "Marañón River Rights Recognized by a Landmark Decision in Peru," Observatoire International des Droits de la Nature, March 21, 2024, https://observatoirenature.org/observatorio/en/2024/03/21/maranon-river-rights-recognized-by-a-landmark-decision-in-peru/.

258. Phil Wilmot, "What's Next for the Struggle to Stop the East Africa Crude Oil Pipeline," *Waging Nonviolence*, April 26, 2024, https://wagingnonviolence.org/2024/04/whats-next-in-struggle-to-stop-eacop-east-africa-crude-oil-pipeline/.

INDEX